**W9-BLQ-781**

β Beta

Multiple-Digit Addition and Subtraction

Instruction Manual

by Steven P. Demme

Math·U·See®

1-888-854-MATH (6284) - mathusee.com
sales@mathusee.com

Beta Instruction Manual: Multiple-Digit Addition and Subtraction

©2012 Math-U-See, Inc.
Published and distributed by Demme Learning

mathusee.com

1-888-854-6284 or +1 717-283-1448 | demmelearning.com
Lancaster, Pennsylvania USA

ISBN 978-1-60826-080-5
Revision Code 1118

Printed in the United States of America by Bindery Associates LLC
 2 3 4 5 6 7 8 9 10

For information regarding CPSIA on this printed material call: 1-888-854-6284
and provide reference #1118-112818

**Building Understanding in Teachers and Students
to Nurture a Lifelong Love of Learning**

At Math-U-See, our goal is to build understanding
for all students.

We believe that education should be relevant, skill-
based, and built on previous learning. Because
students have a variety of learning styles, we
believe education should be multi-sensory. While
some memorization is necessary to learn math
facts and formulas, students also must be able to
apply this knowledge in real-life situations.

Math-U-See is proud to partner with teachers
and parents as we use these principles
of education to **build lifelong learners.**

Curriculum Sequence

$\int$	**Calculus**
$\cos$	**PreCalculus** with Trigonometry
xy	**Algebra 2**
Δ	**Geometry**
x^2	**Algebra 1**
x	**Pre-Algebra**
ζ	**Zeta** Decimals and Percents
ε	**Epsilon** Fractions
δ	**Delta** Division
γ	**Gamma** Multiplication
β	**Beta** Multiple-Digit Addition and Subtraction
α	**Alpha** Single-Digit Addition and Subtraction
P	**Primer** Introducing Math

Math-U-See is a complete, K-12 math curriculum that uses manipulatives to illustrate and teach math concepts. We strive toward "Building Understanding" by using a mastery-based approach suitable for all levels and learning preferences. While each book concentrates on a specific theme, other math topics are introduced where appropriate. Subsequent books continuously review and integrate topics and concepts presented in previous levels.

Where to Start

Because Math-U-See is mastery-based, students may start at any level. We use the Greek alphabet to show the sequence of concepts taught rather than the grade level. Go to mathusee.com for more placement help.

Each level builds on previously learned skills to prepare a solid foundation so the student is then ready to apply these concepts to algebra and other upper-level courses.

Major concepts and skills for Beta:

- Understanding place value and using it to add/subtract
- Fluently adding whole numbers
- Solving for an unknown addend
- Fluently subtracting whole numbers
- Solving abstract and real-world problems involving addition and subtraction

Additional concepts and skills for Beta:

- Telling and writing time by hours and minutes
- Understanding, adding, and subtracting U.S. currency
- Measuring and estimating length with inches, feet, centimeters, and meters
- Comparing numbers and lengths
- Expressing differences between numbers as inequalities
- Finding the perimeter of any polygon
- Representing and interpreting data in plots and graphs

Find more information and products at mathusee.com

Contents

HOW TO USE

Five Minutes for Success

Welcome to Beta. I believe you will have a positive experience with the unique Math-U-See approach to teaching math. These first few pages explain the essence of this methodology, which has worked for thousands of students and teachers. I hope you will take a few minutes and read through these steps carefully.

I am assuming your student has mastered the addition and subtraction facts.

If you are using the program properly and still need additional help, you may visit us online at mathusee.com or call us at 888-854-6284.

–Steve Demme

The Goal of Math-U-See

The underlying assumption or premise of Math-U-See is that the reason we study math is to apply math in everyday situations. Our goal is to help produce confident problem solvers who enjoy the study of math. These are students who learn their math facts, rules, and formulas and are able to use this knowledge to solve word problems and real-life applications. Therefore, the study of math is much more than simply committing to memory a list of facts. It includes memorization, but it also encompasses learning the underlying concepts of math that are critical to successful problem solving.

Support and Resources

Math-U-See has a number of resources to help you in the educational process.

Many of our customer service representatives have been with us for over 10 years. They are able to answer your questions, help you place your student in the appropriate level, and provide knowledgeable support throughout the school year.

Visit mathusee.com to use our many online resources, find out when we will be in your neighborhood, and connect with us on social media.

More than Memorization

Many people confuse memorization with understanding. Once while I was teaching seven junior high students, I asked how many pieces they would each receive if there were fourteen pieces. The students' response was, "What do we do: add, subtract, multiply, or divide?" Knowing **how** to divide is important, but understanding **when** to divide is equally important.

The Suggested 4-Step Math-U-See Approach

In order to train students to be confident problem solvers, here are the four steps that I suggest you use to get the most from the Math-U-See curriculum.

Step 1. Prepare for the lesson
Step 2. Present and explore the new concept together
Step 3. Practice for mastery
Step 4. Progress after mastery

Step 1. Prepare for the lesson

Watch the video lesson to learn the new concept and see how to demonstrate this concept with the manipulatives when applicable. Study the written explanations and examples in the instruction manual.

Step 2. Present and explore the new concept together

Present the new concept to your student. Have the student watch the video lesson with you, if you think it would be helpful. The following should happen interactively.

a. **Build:** Use the manipulatives to demonstrate and model problems from the instruction manual. If you need more examples, use the appropriate lesson practice pages.

b. **Write:** Write down the step-by-step solutions as you work through the problems together, using manipulatives.

c. **Say:** Talk through the why of the math concept as you build and write.

Give as many opportunities for the student to "Build, Write, Say" as necessary until the student fully understands the new concept and can demonstrate it to you confidently. One of the joys of teaching is hearing a student say, *"Now I get it!"* or *"Now I see it!"*

Step 3. Practice for mastery

Using the lesson practice problems from the student workbook, have students practice the new concept until they understand it. It is one thing for students to watch someone else do a problem; it is quite another to do the same problem themselves. Together complete as many of the lesson practice pages as necessary (not all pages may be needed) until the student understands the new concept, demonstrating confident mastery of the skill. Remember, to demonstrate mastery, your student should be able to teach the concept back to you using the Build, Write, Say method. Give special attention to the word problems, which are designed to apply the concept being taught in the lesson. If your student needs more assistance, go to mathusee.com to find review tools and other resources.

Step 4. Progress after mastery

Once mastery of the new concept is demonstrated, advance to the systematic review pages for that lesson. These worksheets review the new material as well as provide practice of the math concepts previously studied. If the student struggles, reteach these concepts to maintain mastery. If students quickly demonstrate mastery, they may not need to complete all of the systematic review pages. You may use the application and enrichment pages for additional practice and for variety.

Now you are ready for the lesson tests. These were designed to be an assessment tool to help determine mastery, but they may also be used as extra worksheets. Your students will be ready for the next lesson only after demonstrating mastery of the new concept and maintaining mastery of concepts found in the systematic review worksheets.

Tell me, I forget. Show me, I understand. Let me do it, I remember.
–Ancient Proverb

To this Math-U-See adds, **"Let me teach it, and I will have achieved mastery!"**

Length of a Lesson

How long should a lesson take? This will vary from student to student and from topic to topic. You may spend a day on a new topic, or you may spend several days. There are so many factors that influence this process that it is impossible to predict the length of time from one lesson to another. I have spent three days on a lesson, and I have also invested three weeks in a lesson. This experience occurred in the same book with the same student. If you move from lesson to lesson too quickly without the student demonstrating mastery, the student will become overwhelmed and discouraged as he or she is exposed to more new material without having learned previous topics. If you move too slowly, the student may become bored and lose interest in math. I believe that as you regularly spend time working along with the student, you will sense the right time to take the lesson test and progress through the book.

By following the four steps outlined above, you will have a much greater opportunity to succeed. Math must be taught sequentially, as it builds line upon line and precept upon precept on previously-learned material. I hope you will try this methodology and move at the student's pace. As you do, I think you will be helping to create a confident problem solver who enjoys the study of math.

Place Value and the Manipulatives

Two skills are needed to function in the decimal system: the ability to count from zero to nine and an understanding of place value. In the decimal system, where everything is based on ten (deci), you count to nine and then start over. To illustrate this, count the following numbers slowly: 800; 900; 1,000. We read these as eight hundred, nine hundred, one thousand. Now read these: 80, 90, 100. Notice how you count from one to nine and then begin again. These are read as eighty, ninety, one hundred. Once you can count to nine, begin work on place value. The two keys are learning the counting numbers zero through nine, which tell us how many; and understanding place value, which tells us what kind.

Counting

When counting, begin with zero and proceed to nine. Traditionally, we've started with one and counted to ten. Look at the two charts that follow and see which is more logical.

1	2	3	4	5	6	7	8	9	10		0	1	2	3	4	5	6	7	8	9
11	12	13	14	15	16	17	18	19	20		10	11	12	13	14	15	16	17	18	19
21	22	23	24	25	26	27	28	29	30		20	21	22	23	24	25	26	27	28	29

The second chart has only single digits in the first line. In the second line, each digit is preceded by a 1 in the tens place. In the next line, a 2 precedes each digit instead of a 1. The first chart, though looking more familiar, has the 10, the 20, and the 30 in the wrong lines. When practicing counting, begin with zero, count to nine, and then start over.

When explaining this important subject, I tell students, "Every value has its own place!" To an older child I would add, "Place determines value!" Both are true. There are ten symbols to tell you how many and many values to represent what kind, or what value. Zero through nine tell us how many; units, tens, and hundreds tell us what kind. For the sake of accuracy, **units** will be used instead of ones to denote the first value. One is a counting number that tells us how many, and units is a place value that denotes what kind. This will save potential confusion when saying ten ones or one ten. Remember, "one" is a number, and "units" is a place value. The numerals 0–9 tell us how many tens, how many hundreds, or how many units. We begin our study focusing on the units, tens, and hundreds, but there are other values such as thousands, millions, billions, and so on.

To illustrate this lesson, I like to use Decimal Street®, since I'm talking about a place. On this street I have the little green units house, the tall blue tens house next door, and the huge red hundreds castle next to the tens. We don't want to forget what we learned from counting—that we count only to nine and then start over. To make this more real, ask, "What is the greatest number of units that can live in this house?" You can get any response to this question, from zero to nine, and you might say "yes" to all of them, but remind the student that the greatest number is nine. We can imagine how many little green beds, or green toothbrushes, or green chairs there would be in the house. Ask the student what else there might be nine of. Do the same with the tens and the hundreds. Remember that in the houses all the furniture will be blue (tens) or red (hundreds).

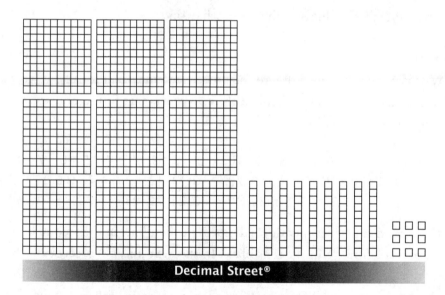

Decimal Street®

When we teach we teach a new concept, we always use the following strategy: Build, Write, Say. To teach place value, we build the number, count how many are in each place, write the number, and read what we've written.

Let's build 142 (1 hundred, 4 tens, 2 units). Now count how many are "at home" at each house. I like to imagine going up to the door of each home and knocking to see how many are at home in each place. Write the numerals 1 4 2 as you count (always beginning with the units) to show the value on paper. Then say, "One hundred, four tens, and two units, or one hundred forty-two." Build another number and have the student write how many are at home. When they understand this, write the number on paper and have them build it. Try 217. After they build it, read what they have built. Keep practicing, going back and forth between the teacher building and the student writing, and vice versa.

Here is another exercise I do to reinforce the fact that every value has its own place. I like to have the student close his/her eyes as I move the pieces around, placing the red hundreds where the units should be, and vice versa. I then ask the student to make sure they are all in the right place. You might call this "scramble the values" or "walk the blocks home." As the student looks at the problem and begins to work on it, I ask, "Is every value in its own place?"

Mr. Zero is a very important symbol. He is a place holder. Let's say you were walking down Decimal Street® and knocking on each door to see who was at home. If you knock on the units door and three units answer, then you have three in the units place. Next door at the tens house, you knock, and no one answers. Yet you know someone is there feeding the goldfish and taking care of the bird, as someone might do when a family goes on vacation. He won't answer the door because he's not a ten, but he's the one who holds their place until the tens come home. Upon knocking at the big red hundreds castle, you find that two hundreds answer the door. Your numeral is thus 203.

Mention that even though we begin at the units end of the street and proceed right to left, from the units to the hundreds, we read the number from left to right. We want to get into the habit of counting units first so that when we add, we will add units first, then tens, then hundreds. On the student worksheets have the students count the correct number at the top of the page; then have them build the number shown next. Remember we teach with the blocks and then move to the worksheets once the student understands the new material.

You've probably noticed the important relationship between language and place value. Consider 142, read as "one hundred forty-two." We know that it is made up of one red hundred square (one hundred), four blue ten bars (forty–*ty* for ten), and two units. The hundreds are very clear and self-explanatory, but the tens are where we need to focus our attention.

When pronouncing 90, 80, 70, 60, and 40, work on enunciating clearly so that 90 is ninety, not "ninedee." 80 is eighty, not "adee." When you pronounce the numbers accurately, not only will your spelling improve, but your understanding of place value will improve as well. Seventy (70) is seven tens, and sixty (60) is six tens. Forty (40) is pronounced correctly but spelled without the *u*. Carrying through on this logic, 50 should be pronounced "five-ty" instead of fifty. Thirty and twenty are similar to fifty, not completely consistent but close enough so we know what they mean.

The teens can be problematic. Some researchers believe that students in Japan and China have a better understanding of place value than students in Europe and the United States. One of the reasons for this is that, in the Chinese and Japanese languages, the words for numbers are very regular, and the words for numbers greater than nine are built quite logically from the words for zero to nine. In contrast, there are a number of irregular words for numbers in English and in other European languages, and the English language in particular is very irregular in the words for eleven through nineteen.

To compensate for this, I'm suggesting a new way to read the numbers 10 through 19. You decide whether this method reinforces the place value concept and restores logic and order to the decimal system. Ten is "onety," 11 is "onety-one," 12 is "onety-two," 13 is "onety-three," and so on. It is not that students can't say ten, eleven, twelve, but learning this method enhances their understanding and makes math logical. Also, children think it is fun.

When presenting place value or any other topic in this curriculum, model how you think as you solve the problems. As you, the teacher, work through a problem with the manipulatives, do it verbally, so that as the student observes, he/she also hears your thinking process. Then record your answer.

Example 1

given visually

Say the number slowly, going from left to right: "Two hundred forty-three." Count, beginning with the units: "1-2-3" and write a 3 in the units place. Count the tens: "1-2-3-4" and write a 4 in the tens place. Finally, count the hundreds: "1-2" and write a 2 in the hundreds place. Do several examples this way and then give the student the opportunity to do some.

Example 2
274 (given as a written number)

Read the number: "Two hundred seventy-four." Say, "Two hundreds" as you pick up two red hundred squares. Then say "Seven-ty or seven tens" and pick up seven blue ten bars. Say "Four" and pick up four green unit pieces. Place the pieces in the correct places as you say, "Every value has its own place." Do several of these problems and then give the student the opportunity to do some.

Example 3
"one hundred sixty-five" (given verbally)

Read the number slowly. Say, "One hundred" as you pick up one red hundred square. Say, "Six-ty or six tens" and pick up six blue ten bars. Now say, "Five" and pick up five green unit pieces. Place the pieces in the correct places as you say, "Every value has its own place." Finally, write the number 165. Do several of these problems and then give the student the opportunity to do some.

Game—Pick a Card

Make up a set of cards with 0 through 9 written in green, with one number on each card. Then make another stack of cards with the same numbers written in blue. Create one more stack of cards with the numbers 0 through 9 written in red. Shuffle the green cards. Then have the child pick a card and display that number of green unit blocks. For example, if a child picks a green 4, four green unit blocks should be counted out.

When the child is proficient at this game, try it with the blue cards and do the same thing except choose the blue ten blocks instead of the green unit blocks. When he can do the tens well, use both sets of cards. Have the child choose one card from the green pile and one card from the blue pile and then choose the correct number of blue ten blocks and green unit blocks. When he or she is an expert at this, add the red cards and proceed as before. Shuffle each set and place them in three stacks. Have the student draw from each stack and show you with the blocks what number was drawn.

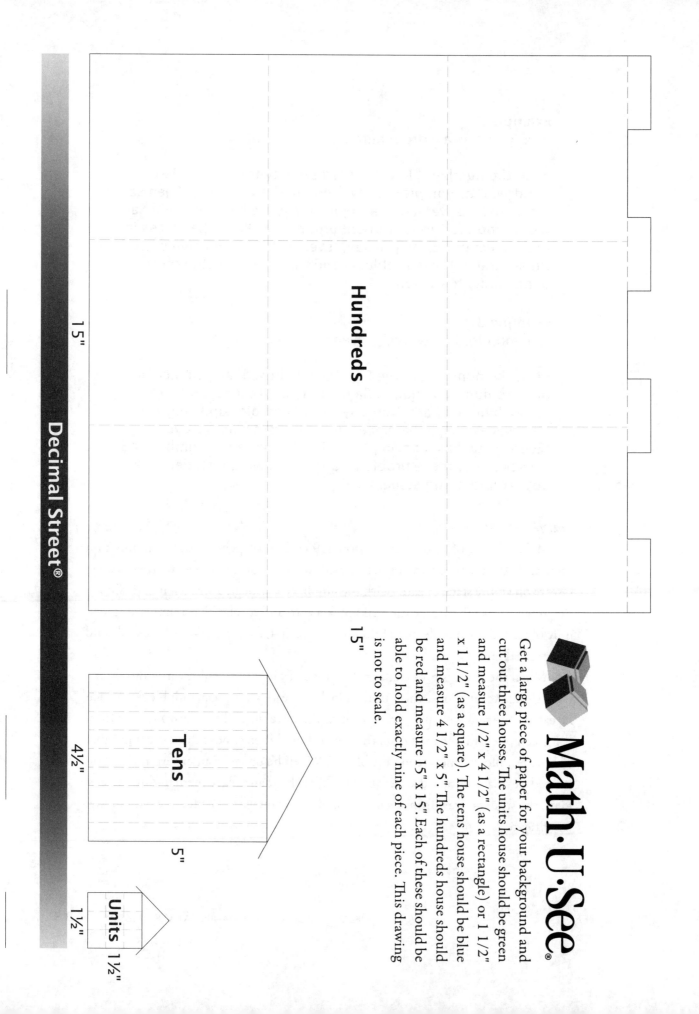

Math·U·See®

Get a large piece of paper for your background and cut out three houses. The units house should be green and measure 1/2" x 4 1/2" (as a rectangle) or 1 1/2" x 1 1/2" (as a square). The tens house should be blue and measure 4 1/2" x 5". The hundreds house should be red and measure 15" x 15". Each of these should be able to hold exactly nine of each piece. This drawing is not to scale.

15"

Hundreds

15"

Tens

4½"   5"

Units

1½"   1½"

Decimal Street®

Sequencing; Word Problem Tips

Sequencing is putting numbers in order from the least to the greatest or the greatest to the least. Using the blocks to represent the numbers before putting them in order makes this exercise very clear. Remember to build, write, and say when doing these problems. The next lesson uses the words and symbols for greater than and less than, while this lesson helps get the student ready to compare numbers by asking the questions, "Which is less?" and "Which is greater?"

This topic also reinforces and reviews place value, which is the cornerstone of the decimal system. Comparing numbers in the same place value is not as difficult, but when comparing and sequencing numbers with different place values, it is imperative to use the blocks to show why the sequence behaves as it does. For example, when comparing numbers such as 95 and 123, it might be helpful to put a 0 in the hundreds place so you are comparing 095 and 123. When comparing numbers with different place values, start with the greatest value, in this case the hundreds, and then move to the right. We know that the greater values are to the left, so start from the greatest place value and move to the least, or from left to right. Emphasize to the student that sequencing is a two-step process. First you have to compare the numbers, and then you have to put them in order. Study the following examples carefully.

When they get proficient at sequencing, we'll do some problems with filling in the missing numbers. (See Examples 5 and 6.) The student may use the blocks for these as well. However, the examples just use blank spaces and numbers.

Example 1
Put numbers in order from least to greatest: 8, 3, 6

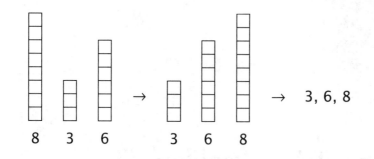 → 3, 6, 8

Example 2
Put numbers in order from least to greatest: 10, 50, 20

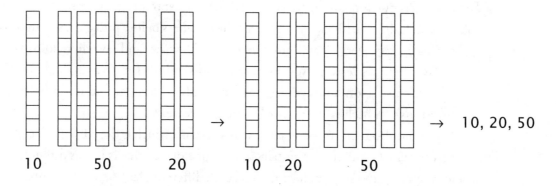 → 10, 20, 50

Example 3
Put numbers in order from greatest to least: 14, 23, 9

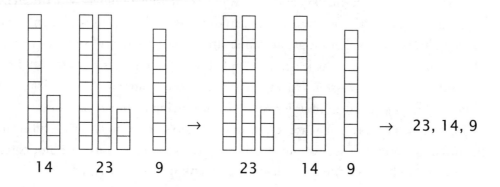 → 23, 14, 9

Example 4
Put numbers in order from greatest to least: 49, 103, 61

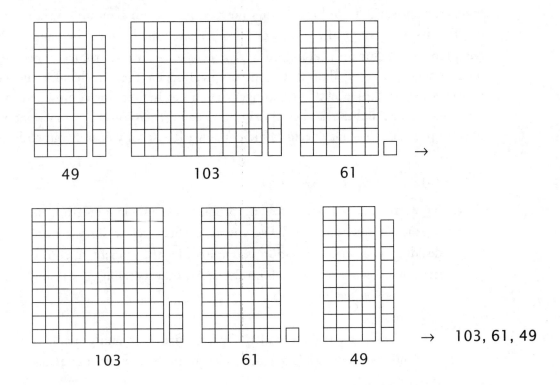

49   103   61

103   61   49   → 103, 61, 49

Example 5
Fill in the blanks with the correct numbers for the sequence.

<u>9</u> , ___ , <u>7</u> , <u>6</u> , ___

Solution: <u>9</u> , <u>8</u> , <u>7</u> , <u>6</u> , <u>5</u>

Example 6
Fill in the blanks with the correct numbers for the sequence.

___ , <u>3</u> , <u>4</u> , ___ , ___

Solution: <u>2</u> , <u>3</u> , <u>4</u> , <u>5</u> , <u>6</u>

Word Problem Tips

Parents often find it challenging to teach children how to solve word problems. Here are some suggestions for helping your student learn this important skill.

The first step is to realize that word problems require both reading and math comprehension. Don't expect a child to be able to solve a word problem if he does not thoroughly understand the math concepts involved. On the other hand, a student may have a math skill level that is stronger than his or her reading comprehension skill. Below are a number of strategies to improve comprehension skills in the context of word problems. You may decide which ones work best for you and your child.

Strategies for word problems:

1. Ignore numbers at first and read the problem. It may help some students to read the question aloud. Every word problem tells a story. Before deciding what math operation is required, let the student retell the story in his own words. Who is involved? Are they receiving gifts, losing something, or dividing a treat?

2. Relate the story to real life, perhaps by using names of family members or friends. For some students, this makes the problem more interesting and relevant.

3. Build, draw, or act out the story. Use the blocks or actual objects when practical. Especially in the lower levels, you may require the student to use the blocks for word problems even when the facts have been learned. Don't be afraid to use a little drama as well. The purpose is to make it as real and meaningful as possible.

4. Look for the common language used in a particular kind of problem. Pay close attention to the word problems on the lesson practice pages, as they model different kinds of language that may be used for the new concept just studied. For example, "altogether" usually indicates addition. These "key words" can be useful clues, but they should not be a substitute for understanding.

5. Look for practical applications that use the concept and ask questions in that context.

6. Have the student invent word problems to illustrate number problems from the lesson.

Cautions:

1. Unneeded information may be included in the problem. For example, we may be told that Suzie is eight years old, but the eight is irrelevant when adding up the number of gifts she received.

2. Some problems may require more than one step to solve. Model these questions carefully.

3. There may be more than one way to solve some problems. Experience will help the student choose the easier or preferred method.

4. Estimation is a valuable tool for checking an answer. If an answer is unreasonable, it is possible that the wrong method was used to solve the problem.

LESSON 3

Inequalities

Read 2 < 3 as "two is less than three." Read 4 > 1 as "four is greater than one." The pointed end of the symbol always points to the lesser number. The open end points to the greater value. Some say the symbol represents an alligator's mouth; the alligator always wants to eat the greater number because it stands for more. Invent your own saying, if you wish. Put the <, >, or = sign in the box to make the math sentence true.

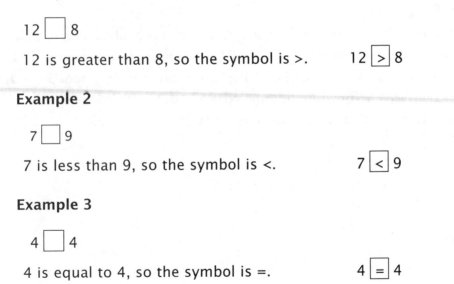

Example 1

12 ☐ 8

12 is greater than 8, so the symbol is >.   12 ⟩ 8

Example 2

7 ☐ 9

7 is less than 9, so the symbol is <.   7 ⟨ 9

Example 3

4 ☐ 4

4 is equal to 4, so the symbol is =.   4 = 4

Example 4

$5 + 7$ ☐ $9 + 4$

12 ☐ 13   Add both sides.

12 is less than 13, so the symbol is <. 12 $\boxed{<}$ 13

Example 5

$10 - 3$ ☐ $9 - 4$

7 ☐ 5   Subtract both sides.

7 is greater than 5, so the symbol is >.   7 $\boxed{>}$ 5

Game — Fill in the Box (Comparison of Amount)

Get a large piece of white paper and draw a vertical line down the middle. Add a box in the center of the dividing line. Make cards with >, <, and =. Put different numbers of blocks on each side of the line and then choose the appropriate card to place in the box.

Facts Review

If your student has mastered single-digit addition and subtraction facts, he or she should be ready to begin *Beta*. We will be reviewing addition facts in the first few lessons. Subtraction is used in solving for the unknown in lessons 5–10, and subtraction facts are systematically reviewed in lessons 11–19. Go to mathusee.com for more resources that may be used to review subtraction facts.

Rounding to 10 and Estimation

Rounding to 10 is used in estimating. When you round a number to the nearest multiple of 10, there will be a digit in the tens place but only a zero in the units place. I tell the students that we call it rounding because the units are going to be a "round" zero.

Let's round 38 as an example. The first skill is to find the two multiples of 10 that are nearest to 38. The lesser one is 30, and the greater one is 40 because thirty-eight is between 30 and 40. If the student has trouble finding these numbers, begin by placing your finger over the 8 in the units place so that all you have is a 3 in the tens place, which is 30. Then add one more to the tens to find the 40. I often write the numbers 30 and 40 above the number 38 on both sides, as in Figure 1.

Figure 1

30   40
   38

The next skill is to find out if 38 is closer to 30 or 40. Let's go through all the numbers given in Figure 2. It is obvious that 31, 32, 33, and 34 are closer to 30 and that 36, 37, 38, and 39 are closer to 40. The number 35 is a special case, since it is just as close to 30 as it is to 40. There are different rules that can be used with numbers ending in 5. The rule we will use here is to round numbers with a 5 in the units place up to the nearest ten. (One reason for this will be explained in lesson 11.) When rounding to tens, look at the units place. If the units are 0, 1, 2, 3, or 4, the digit in the tens place remains unchanged. If the units are 5, 6, 7, 8, or 9, the digit in the tens place increases by one.

Figure 2

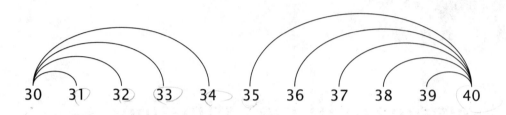

Another strategy I use is to put 0, 1, 2, 3, and 4 inside a circle to represent zero because, if these numbers are in the units place, they add nothing to the tens place. This means they are rounded to the lesser ten (30 in the example). Then I put 5, 6, 7, 8, and 9 inside a thin rectangle to represent one because, if these numbers are in the units place, they do add one to the tens place. This means they are rounded to the greater ten (40 in the example).

Figure 3

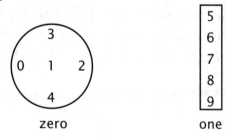

zero   one

Example 1
Round 13 to the nearest ten.

10 20 1. Find the multiples of 10 nearest to 13.
   13

10⌐ 20 2. We see that 3 goes to the lesser ten, 10.
   13

⑩⌐ 20 3. Also, recall that 3 is in the circle, or zero, so
   13   nothing is added to the lesser ten, 10.

Example 2
Round 75 to the nearest ten.

70   80   1. Find the multiples of 10 nearest to 75.
    75

70 ⌐ 80   2. We know that 5 goes to the greater ten, 80.
    75

70 ⌐ (80)   3. Recall that 5 is in the rectangle, or one, so 1
    75   is added to the 7, making the answer 80.

Estimation

Now that we have learned to round numbers, we can apply this knowledge to helping us *estimate*, or find the approximate answer. In the next lesson, we will compare the estimate to the real answer, and if the numbers are close, we are probably correct in our computation.

Example 3

$$
\begin{array}{rl}
13 & (10) \\
+ 75 & (80) \\
\hline
88 & (90)
\end{array}
$$

Round the numbers and put them in parentheses.
Then add the estimates to get 90.

After working the problem exactly, compare your answers to see if they are close.

Commutative Squares

A fun way to review addition facts is to use what some call magic squares. Here is how these work.

1. Add the numbers vertically to find two sums.

2. Add the numbers horizontally to find two more sums.

3. Next, add the sums at the bottom horizontally to find the sum that goes in the bottom right square.

4. Finally, add the sums on the right vertically to find the answer for the bottom right-hand square.

The results of the last two problems should agree in the bottom right square. If not, some of your addition needs to be corrected. Have fun with the examples in the student workbook.

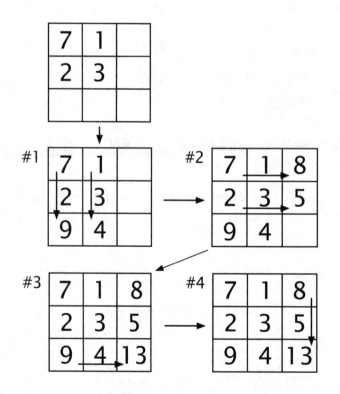

Multiple-Digit Addition
Place-Value Notation

Add the units first. Always add from right to left, that is, from lesser to greater. If this is confusing, simply explain that we read the numbers from left to right, but we combine them from right to left. Begin properly, even though it may not make a difference at this point. Place the four bar and the five bar end to end and place a nine bar next to or on top of it to show that 4 + 5 = 9. Next, push all the tens together so you see that two tens plus three tens is the same as five tens. Finally, add the hundreds. Notice that you add units to units, tens to tens, and hundreds to hundreds. Whenever you add two numbers, you always add the same values. To combine, numbers must be the same kind. Here are three examples followed by the same problems worked out with place-value notation.

Place-Value Notation

Place-value notation is simply writing out the numbers and separating the place values. It follows the format of the blocks. For example, 123 is written as 100 + 20 + 3. This notation reinforces place value.

Example 1
Write 971 with place-value notation.

900 + 70 + 1

Example 2

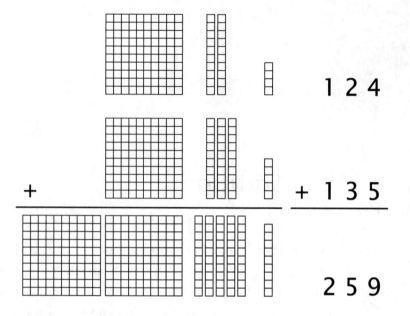

$$124$$

$$+ 135$$

$$259$$

Place-value notation:

$$124 = 100 + 20 + 4$$
$$+135 = 100 + 30 + 5$$
$$259 = 200 + 50 + 9$$

The numbers being added are called the ***addends***. The answer is the ***sum***. In the example above, 124 and 135 are addends, and 259 is the sum.

Example 3

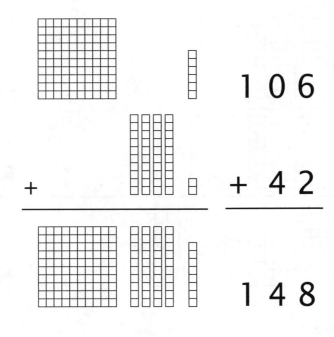

$$
\begin{array}{r}
1\ 0\ 6 \\
+\ 4\ 2 \\
\hline
1\ 4\ 8
\end{array}
$$

Place-value notation:

$$
\begin{array}{rl}
106= & 100\qquad +6 \\
+\ \ 42= & \quad\ \ 40+2 \\
\hline
148= & 100+40+8
\end{array}
$$

Example 4

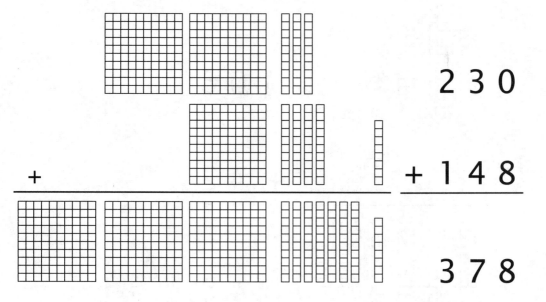

$$230$$
$$+148$$
$$378$$

Place-value notation:

$$230 = 200 + 30$$
$$+148 = 100 + 40 + 8$$
$$378 = 300 + 70 + 8$$

Solving for the Unknown

On Systematic Review 5D in the student workbook, solving for the unknown is reviewed. Throughout the workbook, there are addition and subtraction word problems that require students to solve for the unknown. Remember to give as much help as is needed with these problems.

Example 5

Conner ate three doughnuts. Cody also ate doughnuts. Together the boys ate five doughnuts. How many doughnuts did Cody eat? Solve for the unknown.

$$3 + D = 5 \text{ doughnuts}$$

Think, "three plus what equals five." Recognize that this is the same as "five minus three equals what." The answer is two doughnuts. You can write it as D = 2.

LESSON 6

Skip Count by 2

Skip counting is the ability to count groups of the same number quickly. For example, if you were to skip count by three, you would skip the 1 and the 2 and say, "3," skip the 4 and the 5 and say, "6," and then count "9-12-15-18," etc. Skip counting by seven is 7-14-21-28-35-42-49-56-63-70.

Here are five reasons for learning to skip count.

1. This skill lays a solid foundation for learning the multiplication facts. We can write $3 + 3 + 3 + 3$ as 3×4. If a child can skip count, he could say, "3-6-9-12." Then he could read 3×4 as "3 counted 4 times is 12." As you learn your skip counting facts, you are learning all of the products in the multiplication facts in order. Multiplication is fast adding of the same number, and skip counting illustrates this beautifully. You can think of multiplication as a shortcut to the skip counting process. Consider 3×5. I could skip count 3 five times as 3-6-9-12-15 to come up with the solution. After I learn my facts I can say, "3 counted 5 times is 15." The latter is much faster.

2. Skip counting prepares students to understand multiplication. I had a teacher tell me that her students had successfully memorized their facts but didn't understand the concept. After she had taught them skip counting, they comprehended what they had learned. Multiplication is fast adding of the same number. I can't multiply to solve $1 + 4 + 6$, but I can multiply to solve $4 + 4 + 4$. Skip counting reinforces the concept of multiplication.

3. Multiple counting is helpful as a skill in itself. A pharmacist attending a workshop told me he skip counted pills as they went into the bottles. Another man said he used the same skill for counting inventory at the end of every workday.

4. Skip counting teaches the multiples of a number, which are important when making equivalent fractions to find common denominators. For example, 2/5 = 4/10 = 6/15 = 8/20. The numbers 2-4-6-8 are the multiples of 2, and 5-10-15-20 are the multiples of 5.

5. In the student book, skip counting is reviewed in sequence form, asking students to fill in the blanks: __, __, __, 12, 15, __, __, __, 27, 30. This encourages them to find patterns in math, and patterns are key to understanding this logical and important subject.

One way to learn the skip counting facts is with the *Skip Count and Addition Songbook*. Included is a CD with the skip count songs from the twos to the nines. Children enjoy it, and it has proven very effective.

Another way to teach skip counting is to count normally and then begin to skip some of the numbers. Look at Example 1. Begin by counting each square: 1-2-3-4-5-6 . . . through 20. After this sequence is learned, skip the first number and just count the second number: 2-4-6 . . . through 20. This is skip counting. You may know it as counting by twos.

When first introducing this method, you might try pointing to each square, and, as you count the first number quietly, ask the student to say the second number loudly. Continue this practice, doing it more quietly each time, until you are just pointing to the first block silently while encouraging the student to say the number loudly when you point to the second square. The student sees it, hears it, says it, and can then write the facts 2-4-6 . . . 20 as well.

Point out to the student that the numbers that we say when we count by two are called *even numbers*, while all the other numbers are *odd numbers*. See the teaching tip at the end of this lesson for hands-on ways to enhance the student's understanding of even numbers.

Example 1

Skip count and write the number on the line. Say it out loud as you count and write. Then write the numbers in the spaces provided beneath the figure. By skipping the first space, we come up with the expression "skip counting." The solution follows at the bottom of the page.

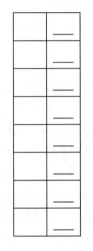

__2__ , __4__ , ___ , ___ , ___ , ___ , ___ , ___

Solution:

	2
	4
	6
	8
	10
	12
	14
	16

__2__ , __4__ , __6__ , __8__ , __10__ , __12__ , __14__ , __16__

Teaching Tip - Even Numbers

Use the blocks to reinforce the idea of even numbers and to show how the numbers that you say when skip counting by two come out even. Count out any even number of unit blocks from 1 to 20. Let's use eight blocks for our example. Give the student the blocks and ask him or her to divide them into two equal shares. There should be four blocks in each share. Ask the student to confirm that the blocks can be divided into two equal or even shares. If each group of blocks is arranged in a line, the student can easily compare the groups to see that each group has the same number of blocks.

Now put the same eight unit blocks back into one pile. Using the 2-bars, match two unit blocks to each 2-bar until all the unit blocks have been used. The student should be able to match up all the unit blocks up evenly with the 2-bars. When the student is finished, skip count by two to find the total. Notice that you get the same answer when you add the shares $(4 + 4 = 8)$ and when you skip count (2-4-6-8).

The student should also be able to predict how many unit blocks will be in each share when they know the total number of blocks. This is based on their knowledge of adding doubles, such as $3 + 3$, $4 + 4$, and $5 + 5$.

Once you have tried this several times with even numbers of unit blocks, try an odd number, such as nine. Challenge the student to divide the blocks into equal shares. They will soon find out why nine is an "odd" number. Follow up by matching the units to the 2-bars as you did before and observe that there is always a leftover (or missing) unit block.

Addition with Regrouping

Don't forget place value—every value has its own place! To take the analogy further, it's okay to visit another place, but there is no place like home! Let's add 34 and 28 to illustrate *regrouping*. First, add the units end to end and look for a ten. You can do this by putting a ten and a two beside the units because 4 + 8 = 10 + 2, or 12. Leave the two units in the units column and "carry" the ten home. (Example 1 is continued on the next page.)

Mr. Ten can visit the units place, but he doesn't live there. I often interject (since I am 6 ft 5 inches myself) that Mr. Ten fell asleep on the couch, and since everything was only nine units long, his feet were hanging over the edge of the couch. We have to carry him "home" to the tens place by picking up the ten bar and placing him with the other tens. We now have 6 tens and 2 units, or 62.

Example 1

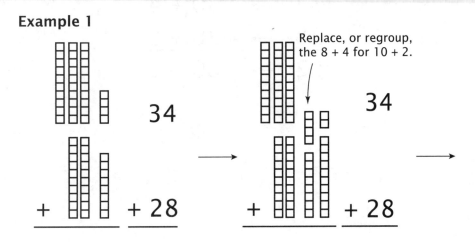

Replace, or regroup, the 8 + 4 for 10 + 2.

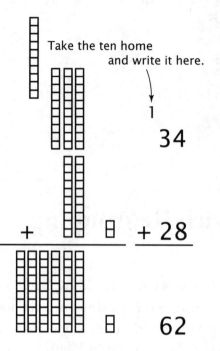

Take the ten home
and write it here.

1
34

+ 28

62

To reinforce the concept of place value, which is the key component in regrouping, I like to use place-value notation to check the work. In the student work, use both methods of checking to make sure the student understands why we regroup the one to the tens place.

$$
\begin{array}{rcl}
\overset{1}{34} & \to & \overset{10}{30} + 4 \\
+\,28 & \to & +\,20 + 8 \\
\hline
62 & \leftrightarrow & 60 + 2
\end{array}
$$

Example 2

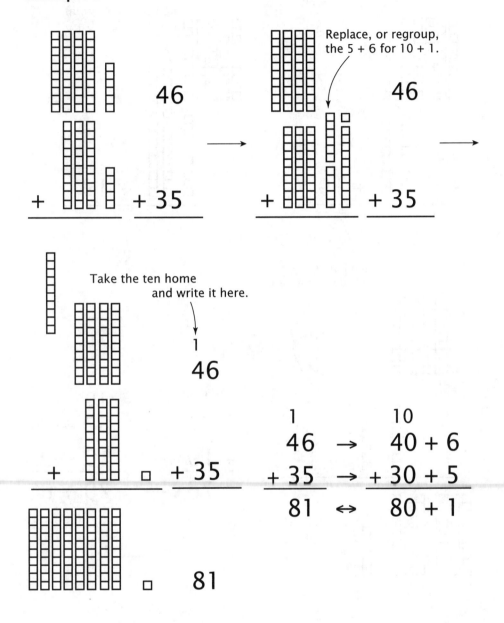

46

+ 35

Replace, or regroup, the 5 + 6 for 10 + 1.

46

+ 35

Take the ten home and write it here.

1
46

+ 35

81

$$
\begin{array}{rcl}
1 & & 10 \\
46 & \rightarrow & 40 + 6 \\
+\ 35 & \rightarrow & +\ 30 + 5 \\
\hline
81 & \leftrightarrow & 80 + 1
\end{array}
$$

Example 3

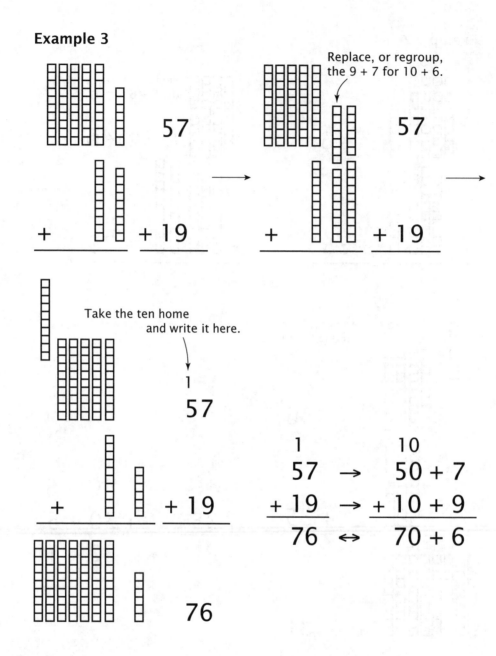

Replace, or regroup, the 9 + 7 for 10 + 6.

57

+ 19

→

57

+ 19

→

Take the ten home and write it here.

1
57

+ 19

76

1		10
57	→	50 + 7
+ 19	→	+ 10 + 9
76	↔	70 + 6

Skip Count by 10
1 Penny = 1¢, 1 Dime = 10¢

After learning to skip count by 2, we need to learn skip counting by 10. Some practical examples are fingers on both hands, toes on both feet, and pennies in a dime. One objective of this lesson is to use skip counting by 10 to find how many pennies equal the value of several dimes. Notice that the multiples of ten (twenty, thirty, forty, etc.) all end in "ty." The suffix *ty* represents ten. Therefore, six-*ty* means six tens, and seven-*ty* means seven tens.

Example 1
As with the twos, skip count and write the number on the line. Say it out loud as you count and write. Then write the numbers in the spaces provided beneath the figure.

									10
									20
									30
									40
									50
									60

___ , ___ , 30 , 40 , ___ , ___

Example 2
Fill in the missing information on the lines.

___ , ___ , ___ , _40_ , ___ , ___ , _70_ , ___ , ___ , ___

Solution
_10_ , _20_ , _30_ , _40_ , _50_ , _60_ , _70_ , _80_ , _90_ , _100_

1 Penny = 1¢

Hold up a penny and teach that this is one penny or one cent. We write it as 1¢. The letter c with a line through it is the symbol used to represent cents. Two cents is 2¢. Nine cents is 9¢. With the blocks we can show one cent by holding up the green unit block.

1¢ = "one penny or one cent" =

Example 1
How many pennies are in 3¢?

   three pennies or 3¢

1 Dime = 10¢

This is a good time to introduce the dime. Dimes are silver-colored and smaller in size than pennies. One dime is equal in value to 10 pennies, even though it is smaller in size than a penny. We represent the dime with the blue ten bar.

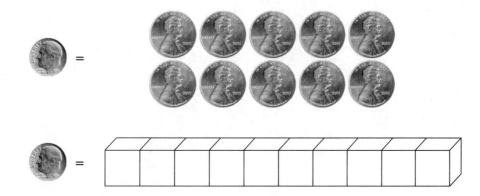

Example 2
How many pennies or cents are in three dimes?

= "10-20-30," so 30 pennies, or 30¢

Example 3
How many pennies or cents are in five dimes?

= "10-20-30-40-50," so 50 pennies, or 50¢

Skip Count by 5
5¢ = 1 Nickel

In this lesson we're teaching skip counting by five. Use the same techniques to introduce and teach this important skill that you have used before. Some practical examples are fingers on one hand, toes on a foot, pennies in a nickel, players on a basketball team, and sides of a pentagon. We can also remind the student that one nickel has the same value as five pennies and apply the skill of counting by fives to find out how many pennies are in several nickels.

Example 1
As with the twos, skip count and write the number on the line. Say the number out loud as you count and write. Then write the numbers in the spaces provided beneath the figure.

				5
				10
				15
				20
				25
				30

___5___ , _____ , _____ , _____ , _____ , _____

If you wish, you may refer to the final number as the *area* of the rectangle. However, be careful not to confuse area and perimeter (taught in lesson 15). Area will be discussed in depth in *Gamma*.

Example 2
Fill in the missing information on the lines.

__5__ , ___ , __15__ , ___ , ___ , __30__ , ___ , __40__ , __45__ , ___

Solution

__5__ , __10__ , __15__ , __20__ , __25__ , __30__ , __35__ , __40__ , __45__ , __50__

5¢ = One Nickel

The nickel is silver-colored and larger in size than the dime or the penny. One nickel is equal in value to five pennies. Using the blocks, we represent the nickel with the light blue five bar because one nickel is five cents, which is the same as five green unit pieces.

Example 3
How many pennies or cents are the same as four nickels?

= "5-10-15-20," so 20 pennies, or 20¢

Example 4

How many pennies or cents are the same as seven nickels?

 = "5-10-15-20-25-30-35," so 35 pennies, or 35¢

Example 5

How many pennies or cents are the same as three nickels?

 = "5-10-15," so 15 pennies, or 15¢

LESSON 10

Money: Decimal Point and Dollars

Working with money reinforces understanding the decimal system since it uses the decimal system exactly, unlike U. S. customary measurement of length (inches, feet, and yards), volume (pints, quarts, and gallons), and weight (ounces, pounds, and tons). Money fits the manipulative blocks as well. The green units represent pennies, the ten bars represent 10 pennies, or one dime, and the hundred squares represent 100 pennies, or one dollar.

To show $2.54, get two hundreds, five tens, and four units. Teach that the $ symbol represents dollars. The decimal point separates the dollars from the cents and is read "and" when reading from left to right. If you have a poster with Decimal Street®, put a black dot between the tens and hundreds for the decimal point. $2.54 is read as "two dollars and fifty-four cents." It may also be read as "two dollars, five dimes, and four pennies." or as "two hundred fifty-four pennies."

Example 1

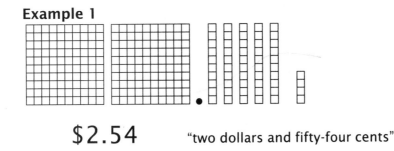

$2.54   "two dollars and fifty-four cents"

Emphasize that 5 dimes are the same as 50 cents, so 5 dimes and 4 pennies are equal to 54 cents. When we read how many cents are to the right of the decimal point, we don't mention the dimes, only the cents.

Example 2

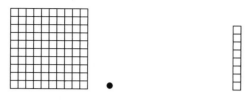

$1.08   "one dollar and eight cents"

Example 3

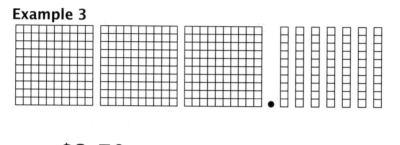

$3.70   "three dollars and seventy cents"

When writing a money amount that does not include whole dollars, it is customary to include a zero in the dollar place. Twenty-five cents would be written as $0.25. Remind the student that the zero means there are no whole dollars.

Rounding to Hundreds
Multiple-Digit Addition with Regrouping

As always, add the units first. A student may become so proficient at adding multiple-digit numbers that he can add from the left, but that method is not appropriate at this stage. For now, always add from right to left, or from least to greatest. Remember that you add units to units, tens to tens, and hundreds to hundreds. Whenever you add two numbers, always add the same values. "To combine, you must be the same kind." The examples show the same problems worked out with place-value notation and regular notation. If your student uses lined paper, I suggest you turn it sideways to help keep the values in the proper places.

Rounding and Estimating to Hundreds

When adding greater numbers, encourage the student to estimate the answer first. We have learned how to round and estimate to the tens place. Now we will increase our understanding by rounding and estimating to the hundreds place.

When you round a number to the nearest multiple of 100, there will be a number in the hundreds place but only zeros in the tens and units places that are to the right of the hundreds place. When rounding, only look at the digit to the immediate right of the place value being considered—in this case, the tens place. This digit determines whether the one in the hundreds place will stay the same or be increased by one. I tell the students we call it rounding because the tens and units are going to be "round" zeros. When we round a number such as 653, we can see one reason why we recommend rounding 5s up instead of down. Although 650 is halfway between 600 and 700, all the other 650s (651, 652, and so on) are closer to 700. Rounding 5s upward makes sense because, if there is another non-zero digit, the number will be closer to the greater number.

When comparing an estimated answer to the final answer, you sometimes see a fairly large difference. For example, 451 will round to 500, and 352 will round to 400. Adding 500 + 400 gives 900, but 451 + 352 is only 803. Remember that an estimate is not intended to be exact. The time to be concerned would be if the final answer were 8,000 or 80.

Example 1
Round 383 to the nearest hundred.

The first step is to find the two multiples of a hundred that are nearest to 383. The lesser one is 300, and the greater one is 400 because 383 is between 300 and 400. If the student has trouble finding these numbers, begin by placing your finger over the 83 so that all you have is a 3 in the hundreds place, which is 300. Then add one more to the hundreds to find the 400. I often write the numbers 300 and 400 above the number 383 on both sides, as shown in Figure 1.

Figure 1
```
300    400
   383
```

Look at the digit in the tens place. Does it fall in 0 through 4 or in 5 through 9? Since it is an 8, it is in the latter group, which means we round up to the next hundred, which is 400. Rounded to the nearest hundred, 383 is 400.

Example 2
Round 547 to the nearest hundred.

```
500    600    1. Find the multiples of one hundred nearest
   547            to 547.
```

```
(500)┐ 600    2. We know that 4 goes to the lesser number,
   547            which is 500.
```

In Examples 3 and 4, the estimates are to the right in parentheses.

Example 3

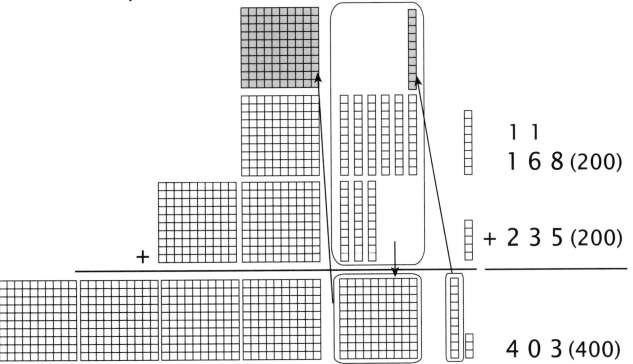

$$\begin{array}{r} 1\ 1 \\ 1\ 6\ 8\ (200) \\ +\ 2\ 3\ 5\ (200) \\ \hline 4\ 0\ 3\ (400) \end{array}$$

Five units plus 8 units equals 13, which is 1 ten and 3 units. We move the ten to the tens place, as indicated by the arrow. Then 6 tens plus 3 tens plus the 1 ten from the result of adding in the units place equals 1 hundred. The 1 hundred is moved to the hundreds place as shown. Example 3 is continued on the next page.

Example 3 (continued)

Adding all the hundreds gives us the answer of 4 hundreds,
0 tens, and 3 units, or 403. The picture below shows the result
after regrouping.

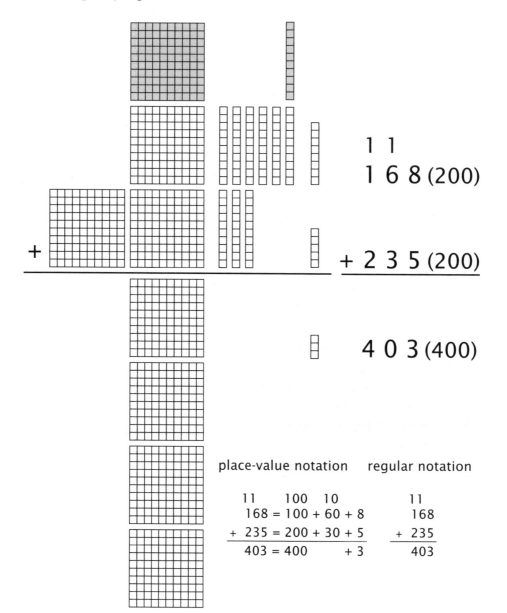

place-value notation   regular notation

```
  11    100  10              11
 168 = 100 + 60 + 8         168
+235 = 200 + 30 + 5       + 235
 403 = 400      + 3         403
```

Example 4

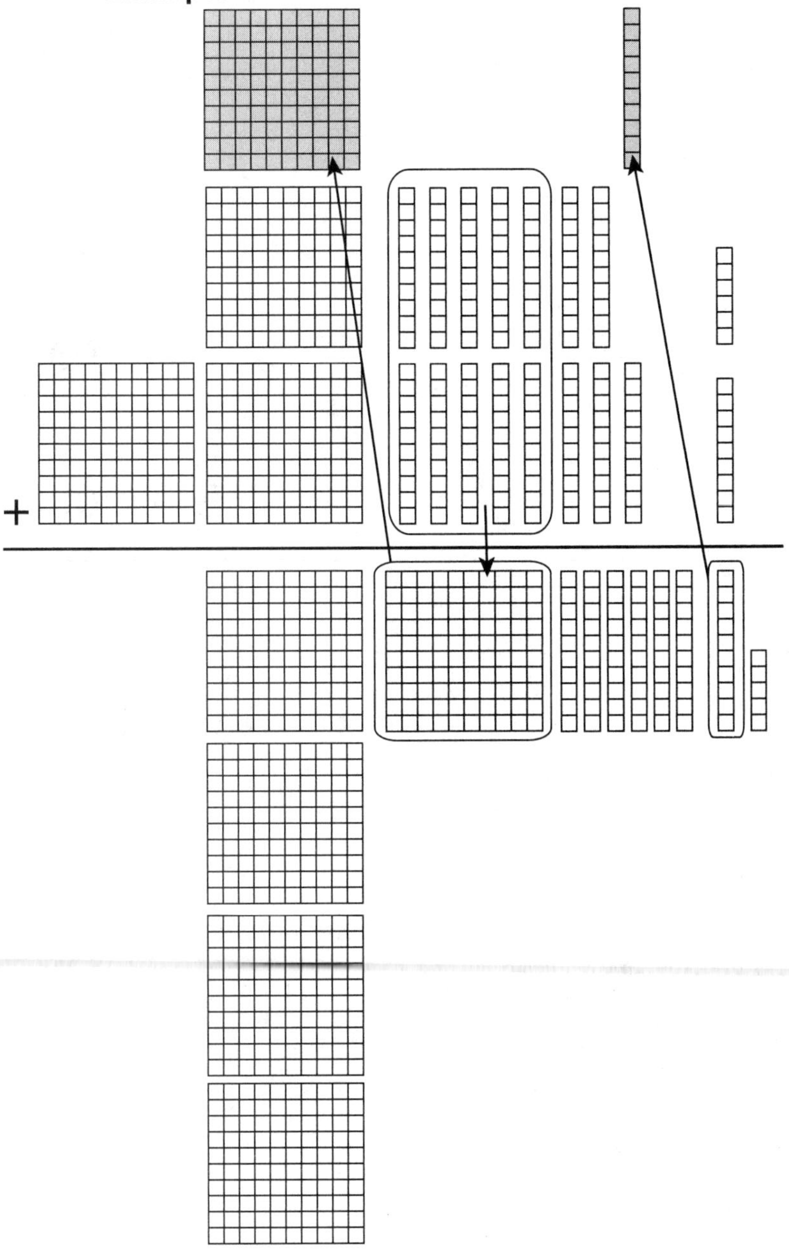

$$
\begin{array}{r}
1\ 1 \\
2\ 7\ 6\ (300) \\
+\ 1\ 8\ 9\ (200) \\
\hline
4\ 6\ 5\ (500)
\end{array}
$$

Example 4 (continued)

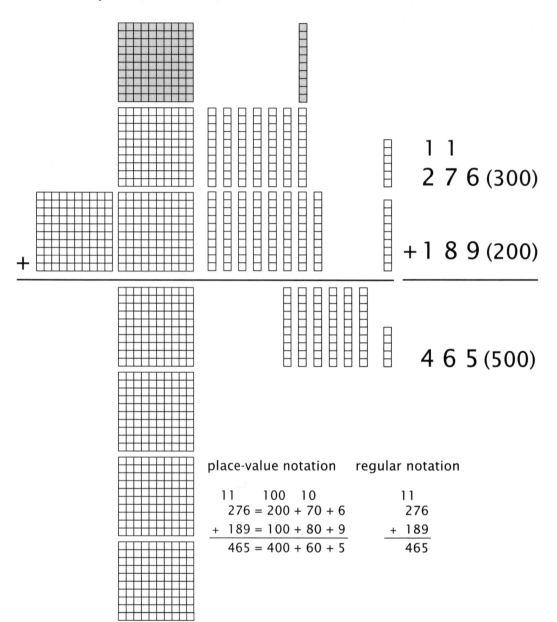

1 1
2 7 6 (300)

+1 8 9 (200)

4 6 5 (500)

place-value notation   regular notation

```
   11    100  10                  11
  276 = 200 + 70 + 6             276
+ 189 = 100 + 80 + 9          +  189
  465 = 400 + 60 + 5             465
```

Adding Money
Mental Math

The red hundreds square represents 100 pennies, or one dollar; the blue ten bar represents ten pennies, or one dime; and the green unit cube represents one penny. Adding money is the same as adding three-digit numbers. If we are adding only pennies, the problem is identical. Since we are adding dollars, dimes, and cents, we need a decimal point. We use the same blocks, but instead of regrouping 10 tens to form one hundred, we are regrouping 10 dimes to make one dollar. Ten pennies are regrouped to make one dime. In the examples, we add the problem two ways: with dollars and decimal points and with pennies.

Emphasize that to combine (add or subtract), things must be the same kind. We add dollars to dollars, dimes to dimes, and cents to cents. The decimal points are lined up when adding so that all the place values, or money values, are aligned.

Example 1 (continued on the next page)

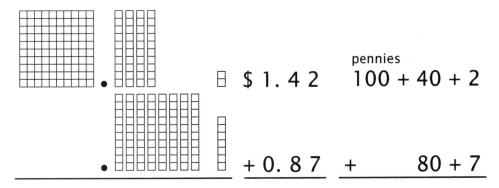

pennies

$1.42   100 + 40 + 2

+ 0.87   +   80 + 7

Solution:

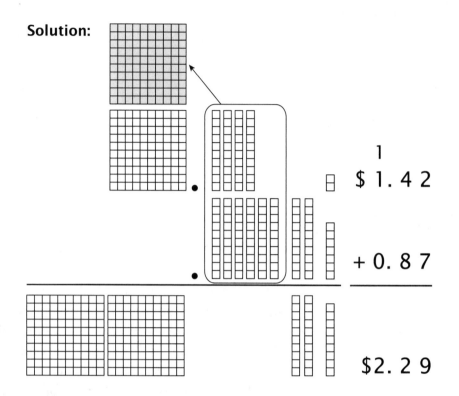

$$\begin{array}{r} 1 \\ \$1.42 \\ +\ 0.87 \\ \hline \$2.29 \end{array}$$

$$\begin{array}{r} \text{pennies} \\ 100 \\ 100 + 40 + 2 \\ +\ 80 + 7 \\ \hline 200 + 20 + 9 \end{array}$$

7¢ plus 2¢ equals 9¢. Eight dimes (or 80¢) plus 4 dimes (or 40¢) is the same as 1 dollar and 2 dimes (or 120¢).

Example 2

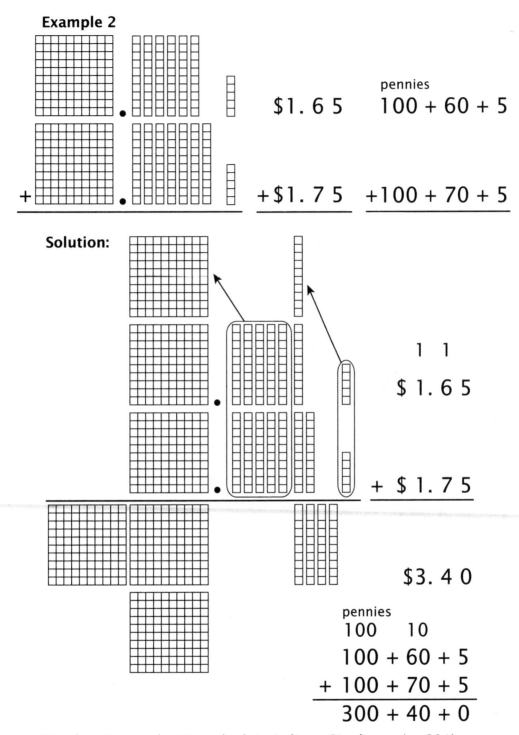

$1. 6 5 $\begin{array}{r} \text{pennies} \\ 100 + 60 + 5 \end{array}$

$+ $1. 7 5 $+100 + 70 + 5$

Solution:

$\begin{array}{r} 1\ \ 1 \\ $1.65 \\ + \ $1.75 \\ \hline \end{array}$

$3. 4 0

pennies
$\begin{array}{r} 100 \quad\ 10 \\ 100 + 60 + 5 \\ + \ 100 + 70 + 5 \\ \hline 300 + 40 + 0 \end{array}$

5¢ plus 5¢ equals 10¢, which is 1 dime. Six dimes (or 60¢) plus 7 dimes (or 70¢), plus the 1 dime from regrouping, is the same as 1 dollar and 4 dimes (or 140¢).

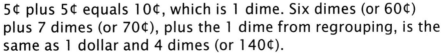

Mental Math

These problems can be used to keep the facts alive in the memory and to develop mental math skills. The teacher should say the problem slowly enough so that the student comprehends and then walk him through increasingly-difficult exercises. The purpose is to stretch but not discourage.

Example 3
Two plus three plus one equals what number?

The student thinks, "Two plus three equals five, and five plus one equals six." Go slowly at first so the student can verbalize the intermediate step. As skills increase, he or she should be able to give just the answer.

Starting with this lesson, selected lessons in the instruction manual will have mental math problems for you to read aloud to your student. Look for them in lessons 18, 21, 24, and 27 and try a few at a time.

1. Seven plus two, plus eight equals what number? (17)

2. One plus two, plus four equals what number? (7)

3. Five plus three, plus two equals what number? (10)

4. Six plus three, plus six equals what number? (15)

5. Eight plus zero, plus five equals what number? (13)

6. Three plus four, plus one, plus three equals what number? (11)

7. Two plus one, plus four, plus five equals what number? (12)

8. One plus two, plus one, plus two equals what number? (6)

9. Four plus one, plus two, plus nine equals what number? (16)

10. One plus five, plus three, plus three equals what number? (12)

Column Addition

Suppose you wanted to add a list of numbers: $3 + 4 + 6 + 5 + 7 = 25$. The key is finding 10 in the list of numbers. In this case, $3 + 7$ and $4 + 6$ make two tens. Adding the two tens and the five units equals 25.

Try this one: $1 + 2 + 3 + 4 + 5 + 6 + 7 + 8 + 9$. We have $\underline{1 + 9}$, $\underline{2 + 8}$, $\underline{3 + 7}$, $\underline{4 + 6}$, $+ 5$, which is four tens and five units, or 45. This is a good way to reinforce regrouping.

It is generally easiest to solve these problems by writing them in a column, as shown in Example 1.

Example 1

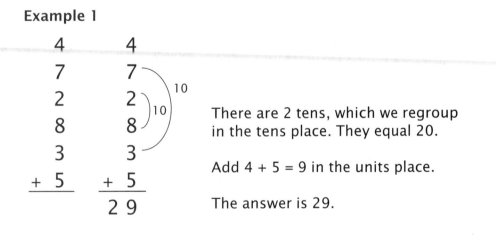

There are 2 tens, which we regroup in the tens place. They equal 20.

Add $4 + 5 = 9$ in the units place.

The answer is 29.

Example 2

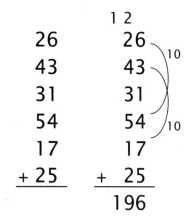

```
          1 2
  26      26
  43      43     10
  31      31
  54      54   10
  17      17
+ 25    + 25
        ─────
         196
```

There are 2 tens, which we regroup in the tens place. They equal 20.

Add 1 + 5 = 6 in the units place.

There are now 19 tens, or 190, so we regroup 1 hundred in the hundreds place and have 9 tens in the tens place.

The answer is 196.

Example 3

```
            3 1
  30        30
  24    100/ 24
  56         56  )10
  74    100⟨ 74
  83    100  83
+ 52       + 52
          ─────
           319
```

There is 1 ten, which we regroup in the tens place.

Add 2 + 3 + 4 = 9 in the units place.

There are now 31 tens, or 310, so we regroup 3 hundreds in the hundreds place and have 1 ten in the tens place.

The answer is 319.

Making tens can simplify column addition, but it will not work with all problems. Show the student that any two numbers in a column can be combined. Write the results of each combination to the side and then add them, combining tens if possible. The numbers in each column may also be added one after another as when doing a mental math problem.

When the student has mastered column addition, we suggest that you go back to lessons 6 and 9 and look at the rectangles that you used for skip counting. Point out that, instead of skip counting, the student can get the same answer by adding. Skip counting 5-10-15-20 gives the same result as 5 + 5 + 5 + 5 = 20. The result tells us the number of blocks that make up the rectangle. If you wish, you may use the word *area* to describe the answer. Area is taught in detail in *Gamma*.

Measurement: 12 Inches = 1 Foot

A symbol for inches is " after the number. Another way to indicate inches is by writing just the first two letters (*in*). Eight inches is written as 8 *in* or 8". A symbol for feet is ' after the number and is abbreviated as *ft*. Six feet is written as *6 ft* or *6'*.

Give the student a 12" ruler. The student may also use a note card to make a ruler by placing it beside the ruler on this page and creating his or her own portable measuring instrument.

Each of the spaces between the vertical lines is one inch long. The whole line is five inches long. One foot is 12 inches long. When I don't have a ruler handy, I use my knuckle to estimate one inch. Have the student measure the distance between two joints on each of his fingers to see which one is the closest to one inch.

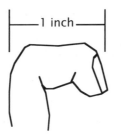

Example 1
Measure the length of the line and write your answer using the inch symbol.

_____

Solution:

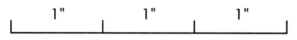

The line is 3" long.

Example 2
Using your ruler, note card, or knuckle, measure the length and width of this book. Give the answer to the nearest inch.

Solution: It should be a little more than 8 inches wide and about 11 inches long, or 8" by 11" rounded to the nearest inch.

Example 3
How many inches are in five feet?

Solution: There are 12 inches in 1 foot. 12 inches + 12 inches + 12 inches + 12 inches + 12 inches equals 60 inches. There are 60 inches in 5 feet.

Using Different Units; Centimeters and Meters

As you measure different objects, discuss how to choose appropriate units. For example, inches are best used to measure the length of a pencil, while feet are a better choice for the dimensions of a room. Also discuss whether a ruler, a yardstick, a tape measure, or some other tool is best for a particular measuring job. Encourage students to estimate the length of an object before measuring it. The student book includes word problems involving addition and subtraction of measures.

Metric measures are taught in detail in the _Zeta_ level of Math-U-See, along with operations using decimal numbers. We do recommend introducing metric measures at this level if the student is comfortable with feet and inches. Most rulers used in the U. S. include centimeters as well as inches. One hundred centimeters make a meter, which is a little longer than a yard. Experience is the best tool for learning about measures, so be sure to give students lots of hands-on experience.

Perimeter

Often older students confuse *perimeter* and area because these terms have been taught simultaneously. One strategy to help them remember the difference is to see the word RIM in peRIMeter. When we ask for perimeter, we are asking for the distance around a shape, or along its RIM. Perimeter comes from the Greek words *meter*, meaning "measure," and *peri*, meaning "around." I teach this concept having the students imagine themselves to be ants crawling around the entire outside edge of a rectangle. Then I ask, "How many units is it around the entire rectangle?" Another strategy is to have the student(s) walk around the perimeter of a room, yard, or field. This helps a student experience the concept of "walking around," rather than just hearing the information.

Perimeter is measured in linear units of measure (inches, feet, yards, etc.). If the unit of measure is not given, label the answer as "units."

Example 1
Find the perimeter of the rectangle.

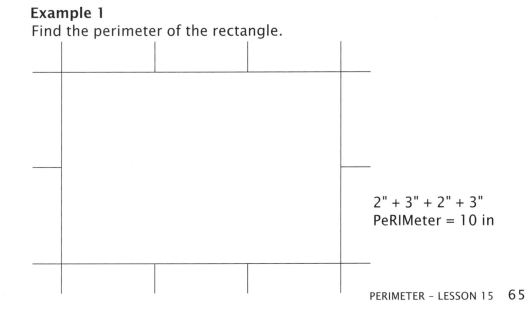

2" + 3" + 2" + 3"
PeRIMeter = 10 in

Example 2
Find the perimeter of the rectangle.

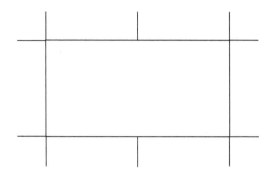

2 in + 1 in + 2 in + 1 in
PeRIMeter = 6 in

Example 3
Find the perimeter of the triangle.

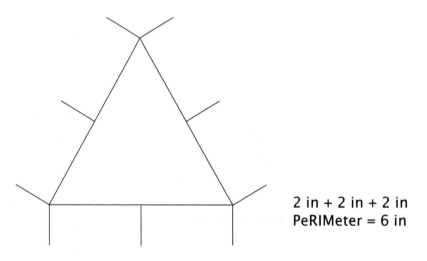

2 in + 2 in + 2 in
PeRIMeter = 6 in

Most students are already able to identify squares, rectangles, and triangles. There is a review of these shapes and their names on 13D in the student book. For more on shapes, go to Appendix A at the end of this book. Fractional parts of shapes are also discussed in the appendix.

Thousands and Place-Value Notation

A huge component in understanding multiple-digit addition is *place value*. The beginning value is units and is represented by the small green unit cubes. The next greatest place value is the tens place, represented by the blue ten bars. A blue bar is 10 times as large as a unit block. The next value as we move to the left is the hundreds, represented by the large red blocks. Each red block is 10 times as large as a ten bar. Notice that as you move to the left, each value is 10 times as great as the preceding value. See Figure 1.

Figure 1

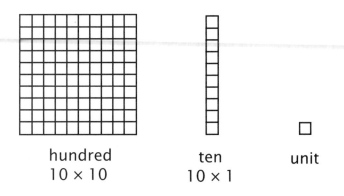

hundred
10 × 10

ten
10 × 1

unit

When you name a number such as 247, the 2 tells you how many hundreds, the 4 indicates how many tens, and the 7 indicates how many units. We read 247 as "two hundred four-ty seven." (The "ty" in forty means ten.) The 2, 4, and 7 are digits that tell us *how many*. The hundreds, tens and units tell us *what kind*, or *what value*. The *place* where the digit is written tells us what *value* it is.

Because we are operating in the base 10 or decimal system, the values increase by a factor of ten each time you move one place to the left. The next place value is the thousands place. It is 10 times 100. You can build 1,000 by stacking 10 hundred squares to make a cube. You can also show 1,000 by making the cube into a rectangle 10 by 100, as in Figure 2. (Notice that the drawing is smaller than actual size.) Ten thousand is also shown in the figure. Can you imagine what 100,000 would look like if you stick to rectangles? It would be a rectangle 100 by 1,000. The factors are written inside the rectangles.

Figure 2

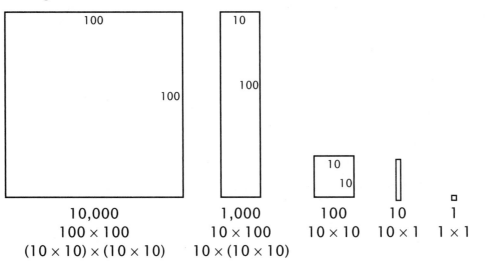

10,000	1,000	100	10	1
100×100	10×100	10×10	10×1	1×1
$(10 \times 10) \times (10 \times 10)$	$10 \times (10 \times 10)$			

Notice that the digits are divided into groups of three. A comma is used to separate groups.

Figure 3

When saying greater numbers, I like to think of the commas as having names. The first comma from the right is named "thousand."

Example 1
Say 456,789.

"456 thousand, 789" or "four hundred fifty-six thousand, seven hundred eighty-nine"

Notice that you never say "and" when reading a greater number. It is reserved for the decimal in decimal numbers. Do you see that you never read a number greater than hundreds between the commas? This is because there are only three places between the commas. Practice saying and writing greater numbers.

Example 2
Say 13,762.

"13 thousand, 762" or "thirteen thousand, seven hundred sixty-two"

More Place-Value Notation
Place-value notation is a way of writing numbers that emphasizes the place value of each digit. The number in Example 2 looks like this in place-value notation: 10,000 + 3,000 + 700 + 60 + 2. Each value is separated from the others.

Note to Teacher
Place-value notation is used primarily to reinforce the role that place value plays in regrouping. Once a student understands this relationship, the continued use of place-value notation may prove confusing. Even though the method is included in the instruction manual, you don't have to continue using it with students once they have mastered the concept.

Example 3
Write 543,971 with place-value notation.

500,000 + 40,000 + 3,000 + 900 + 70 + 1

Counting by 100s
This is a good time to practice counting by 100s to 1,000. Start at 100, or at any multiple of 100, and count on from there.

Example 4
Start at 100 and count by 100s to 1,000.

100, 200, 300, 400, 500, 600, 700, 800, 900, 1,000

Rounding to Thousands
Estimation

Most of this lesson should be review, since we have covered this material in previous lessons. Take your time to digest the new material thoroughly. Rounding is used in estimating as we add and subtract.

When you round a number to the nearest multiple of 1,000, there will be a digit between one and nine in the thousands place and zeros in the places to the right of the thousands place. The digit to the right of the place value to which we are rounding determines whether to stay the same or increase by one. Perhaps we call it rounding because the hundreds, tens, and units are going to be "round" zeros.

Example 1
Round 4,299 to the nearest thousand.

The first step is to find the two multiples of a thousand that are nearest to 4,299. The lesser one is 4,000, and the greater one is 5,000. The number 4,299 is between 4,000 and 5,000.

Look at the digit in the hundreds place. Does it fall between 0–4 or 5–9? Since it is a 2, it is in the first group, which means the digit in the thousands place stays the same, and the other digits are "rounded" to zero. Thus, 4,299 rounded to the nearest thousand is 4,000.

4,000　5,000 4,299	1. Find the multiples of 1,000 nearest to 4,299.
(4,000) 5,000 4,299	2. We know that 2 stays the same, so the answer is 4,000.

Example 2
Round 6,502 to the nearest thousand.

6,000   7,000   1. Find the multiples of 1,000 nearest to
    6,502   6,502.

6,000 ⌐(7,000)   2. We know that 5 increases the
    6,502   thousands place by 1 to 7,000.

When rounding to thousands, remember to look only at the digit in the hundreds place to determine whether to stay the same or increase by one. The same rules apply to thousands that apply to hundreds and tens: if the digit in the hundreds place is a 0, 1, 2, 3, or 4, the number in the thousands place remains unchanged. If the digit is a 5, 6, 7, 8, or 9, then the number in the thousands place increases by one.

When rounding to ten thousands, hundred thousands, or millions, consider only the digit immediately to the right of the place to which you are rounding.

Example 3
Round 27,601 to the nearest ten thousand.

20,000   (30,000)   1. Find the multiples of 10,000
    27,601   nearest to 27,601.

20,000 ⌐ (30,000)   2. The 7 to the right of 2 increases
    27,601   the ten thousands place by 1 to
                                30,000.

Estimation

The main reason for rounding is to help us estimate an answer. In Example 4, we'll use the results of Examples 1 and 2 to estimate the answer to an addition problem. The results of rounding and the estimated answer are in parentheses. After adding the original numbers, compare the answer to the estimate. The two should be fairly close in value.

Example 4

$$
\begin{array}{ll}
4{,}299 \ (4{,}000) & 4{,}299 \ (4{,}000) \\
+\ 6{,}502 \ (7{,}000) \rightarrow & +\ 6{,}502 \ (7{,}000) \\
\hline
\quad\ (11{,}000) & 10{,}801 \ (11{,}000)
\end{array}
$$

In Example 5, we'll use estimation to check our answer to a subtraction problem. (This skill will be used in later lessons.) The results of rounding and the estimated answer are in parentheses.

Example 5
Round 7,874 and 5,120 to the nearest thousand.

$$
\begin{array}{ll}
7{,}000 \ \lceil \boxed{8{,}000} & \boxed{5{,}000} \ \rceil\ 6{,}000 \\
\quad 7{,}874 & \quad 5{,}120
\end{array}
$$

$$
\begin{array}{ll}
7{,}874 \ (8{,}000) & 7{,}874 \ (8{,}000) \\
-\ 5{,}120 \ (5{,}000) \rightarrow & -\ 5{,}120 \ (5{,}000) \\
\hline
\quad\ (3{,}000) & 2{,}754 \ (3{,}000)
\end{array}
$$

Multiple-Digit Column Addition
Mental Math

Now we get a chance to use everything we have learned so far in this book. Place value, addition facts, regrouping, column addition, and making 10 all play a role in these problems. Look for ways to make the big problems into smaller ones. Here are a few examples to help you. Try them yourself first and then compare your work to the solution shown. Don't forget to estimate your answer first. The estimates are shown in the parentheses.

The place value to the left of the hundreds is the thousands. One of the main problems that occurs is keeping the values in the proper places. If you are using lined paper, consider turning your paper sideways to make the best use of the lines for lining up place value.

Example 1

```
       2 2 2
 758    758    (800)
 342    342    (300)
 167    167    (200)
 532    532    (500)
+956   +956  (1,000)
       2,755  (2,800)
```

Use making 10 as much as possible in each column to make your work easier.

Adding the units column gives us 25. Put 2 tens with the other tens and leave 5 in the units column.

Adding the tens column gives us 25, which is really 250 because we are adding tens and not units. Put 2 in the hundreds column and leave 5 in the tens column.

Adding the hundreds column gives us 27, which is really 2,700. Put 2 in the thousands column and leave 7 in the hundreds column.

Example 2

```
      1 2 2
263   263   (300)
 47    47  *(0)
259   259   (300)
558   558   (600)
+310  +310  (300)
     1,437 (1,500)
```

Adding the units column gives us 27. Put 2 tens with the other tens and leave 7 in the units column.

Adding the tens column gives us 23, which is really 230 because we are adding tens and not units. Put 2 in the hundreds column and leave 3 in the tens column.

Adding the hundreds column gives us 14, which is really 1,400. Put 1 in the thousands column and leave 4 in the hundreds column.

Example 3

```
      2 2 1
527   527   (500)
 86    86  *(100)
364   364   (400)
411   411   (400)
+690  +690  (700)
     2,078 (2,100)
```

Adding the units column gives us 18. Put 1 ten with the other tens and leave 8 in the units column.

Adding the tens column gives us 27, which is really 270 because we are adding tens and not units. Put 2 in the hundreds column and leave 7 in the tens column.

Adding the hundreds column gives us 20, which is really 2,000. Put 2 in the thousands column and leave 0 in the hundreds column.

*rounding to hundreds

Mental Math

Here are some more questions to read to your student. These review subtraction. Remember to go slowly at first.

1. Nine minus one, minus four equals what number? (4)

2. Eight minus six, minus zero equals what number? (2)

3. Ten minus five, minus two equals what number? (3)

4. Sixteen minus nine, minus one equals what number? (6)

5. Fourteen minus five, minus four equals what number? (5)

6. Six minus three, minus one equals what number? (2)

7. Eleven minus seven, minus three equals what number? (1)

8. Eighteen minus nine, minus four equals what number? (5)

9. Fifteen minus one, minus six equals what number? (8)

10. Ten minus two, minus two equals what number? (6)

More Multiple-Digit Column Addition

The purpose of this lesson is to offer more practice. You decide how much time the student needs to spend on this. The skills already taught are sufficient to do any addition problem. Teach students to take their time, put the values in the proper places, and make 10. If you would like, after your student feels comfortable solving these problems, you can let him check his work on a calculator.

In our estimations, we will be rounding to thousands, so everything to the right of the thousands place will be zeros. Answers may include the ten thousands place.

Example 1

```
          11 11
  8,534    8,534    (9,000)
  2,761    2,761    (3,000)
 +3,659   +3,659    (4,000)
          14,954   (16,000)
```

Adding the units column gives us 14. Put 1 ten with the other tens and leave 4 in the units column.

Adding the tens column gives us 15, which is really 150 because we are adding tens and not units. Put 1 in the hundreds column and leave 5 in the tens column.

Adding the hundreds column gives us 19, which is really 1,900. Put 1 in the thousands column and leave 9 in the hundreds column.

Adding the thousands column gives us 14, which is really 14,000. Put 1 in the ten thousands column and leave 4 in the thousands column.

Example 2

```
          2 1  1
 3,742     3,742    (4,000)
 9,555     9,555   (10,000)
+8,310    +8,310    (8,000)
          ──────
          21,607   (22,000)
```

Adding the units column gives us 7. There are no tens to regroup. Leave 7 in the units column.

Adding the tens column gives us 10, which is really 100 because we are adding tens and not units. Put 1 in the hundreds column and leave 0 in the tens column.

Adding the hundreds column gives us 16, which is really 1,600. Put 1 in the thousands column and leave 6 in the hundreds column.

Adding the thousands column gives us 21, which is really 21,000. Put 2 in the ten thousands column and leave 1 in the thousands column.

Example 3

```
         1 1
 7,068    7,068    (7,000)
   460      460      (0)
+  37    +   37      (0)
         ──────
          7,565    (7,000)
```

Adding the units column gives us 15. Put 1 ten with the other tens and leave 5 in the units column.

Adding the tens column gives us 16, which is really 160 because we are adding tens and not units. Put 1 in the hundreds column and leave 6 in the tens column.

Adding the hundreds column gives us 5, which is really 500. There are no thousands to regroup. Leave 5 in the hundreds column.

Adding the thousands column gives us 7, which is really 7,000. There are no ten thousands to regroup. Leave 7 in the thousands column.

LESSON 20

Multiple-Digit Subtraction

We have always said that to compare or combine you must have the same kind. This rule applies to subtraction as well as addition. When subtracting multiple-digit numbers, subtract from right to left. In other words, subtract units from units, tens from tens, and hundreds from hundreds. In the examples, place-value notation is used as well as regular notation to emphasize this principle.

If using lined paper, the student can turn it sideways and use the lines to keep the place values straight.

Example 1

$$
\begin{array}{r}
957 \\
-342 \\
\hline
615
\end{array}
\qquad
\begin{array}{r}
900+50+7 \\
-300+40+2 \\
\hline
600+10+5
\end{array}
$$

Example 2

$$
\begin{array}{r}
598 \\
-208 \\
\hline
390
\end{array}
\qquad
\begin{array}{r}
500+90+8 \\
-200+00+8 \\
\hline
300+90+0
\end{array}
$$

Example 3

$$
\begin{array}{r}
764 \\
-\ 62 \\
\hline
702
\end{array}
\qquad
\begin{array}{r}
700+60+4 \\
-\quad\ 60+2 \\
\hline
700+00+2
\end{array}
$$

Example 4

$$\begin{array}{r} 253 \\ -120 \\ \hline 133 \end{array} \qquad \begin{array}{r} 200+50+3 \\ -100+20+0 \\ \hline 100+30+3 \end{array}$$

In a subtraction problem, the top number is the *minuend,* the second number is the *subtrahend*, and the answer is the *difference*. You can easily check the answer to a subtraction problem by adding the subtrahend and the difference.

From Example 1

$$\begin{array}{r} 957 \\ -342 \\ \hline 615 \end{array} \qquad \begin{array}{r} 957 \\ -342 \\ \hline 615 \\ \hline 957 \end{array}$$

To check this answer, I like to draw a wavy line under the 615 and then add 342 and 615. This sum should equal 957. The arrow shows that the two numbers match.

Consider having students check their answers this way on the worksheets. If the wavy line and arrow prove to be confusing, have them rewrite the problem beside the original as demonstrated below.

From Example 1

$$\begin{array}{r} 957 \\ -342 \\ \hline 615 \end{array} \qquad \begin{array}{r} 342 \\ +615 \\ \hline 957 \end{array}$$

LESSON 21

Telling Time: Minutes
Mental Math

When students have mastered skip counting by five, they are ready to learn how to tell time. This can be a challenge with a clock that is not digital. We'll begin by taking 6 ten bars and explaining that there are 60 minutes in one hour. Next replace each ten bar with 2 five bars. If you don't have enough five bars, you can make this clock by using various combinations for five; for example, five units, or a two and a three, or a four and a unit. Take your 12 groups of five and arrange them in a circle (really a dodecagon, or 12-sided polygon) so you have 60 minutes in the shape of a clock. Beginning at the top, skip count by five going around the clock: 5-10-15 and up to 60. Since 60 minutes is one hour, the number of minutes goes back to zero, and we start over again.

Choose a long unit bar, turn it on its smooth side, and use it for a minute hand. Point to a time and, beginning at the top of the clock, count the minutes to that time. Even though we emphasize counting by five for reading the clock and finding the minutes, remember that there are 60 possible minutes that are not all multiples of five. When giving problems on your own block clock, move the minute hand to positions signifying 1, 2, 8, 34, or 51.

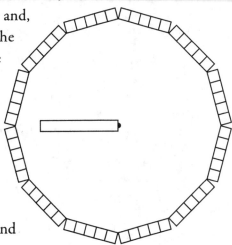

There is a template for your clock at the end of the student workbook.

To help the students see the progression of the minutes, build several partial clocks as in Examples 1 and 2. Count the minutes by beginning at the top and moving around to the right, or clockwise.

Example 1
Count the minutes. 5-10; 10 minutes

Example 2
Count the minutes. 5-10-15; 15 minutes

Mental Math

Here are some more questions to read to your student. These combine addition and subtraction. Don't try the longer problems unless the student is comfortable with the shorter ones.

1. Four plus five, minus three equals what number? (6)

2. Seven minus two, plus six equals what number? (11)

3. Two plus nine, minus nine equals what number? (2)

4. Five minus three, plus ten equals what number? (12)

5. Ten minus eight, plus five equals what number? (7)

6. Seven minus six, plus eight, minus four equals what number? (5)

7. Four plus seven, minus three, minus seven equals what number? (1)

8. Eight plus eight, minus six, plus one equals what number? (11)

9. Ten minus two, plus four, minus six equals what number? (6)

10. Fourteen minus eight, plus two, minus four equals what number? (4)

LESSON 22

Subtraction with Regrouping

Now that the student is comfortable adding and regrouping, we can apply this knowledge to the inverse: subtracting and regrouping. The following examples illustrate the process. When we are not able to subtract, we regroup from the next greatest place value.

Example 1

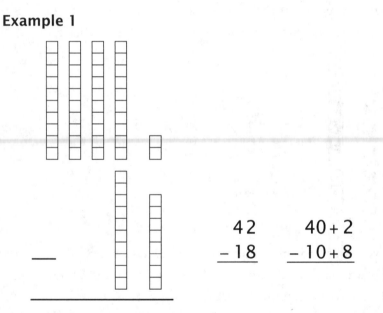

$$
\begin{array}{r}
42 \\
-18 \\
\hline
\end{array}
\qquad
\begin{array}{r}
40+2 \\
-10+8 \\
\hline
\end{array}
$$

The question is: "Can we count 8 from 2?" No, two isn't enough to have 8 counted from it. The number on top has to be the same or greater than the one underneath it. Ask the student if he has ever run to the neighbor's house to borrow a cup of flour, or milk, or whatever. Since there are not enough units, we need to go to our neighbor (the tens) and "borrow" one ten. (Continued on the next page.)

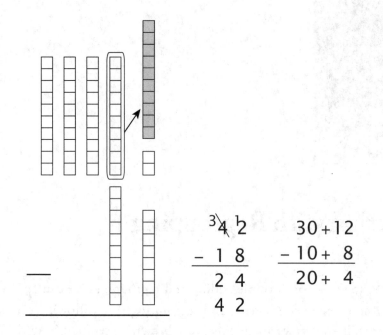

$$\begin{array}{r} {}^3\!\!\not{4}\;{}^1\!\!2 \\ -\;1\;8 \\ \hline 2\;4 \\ 4\;2 \end{array} \qquad \begin{array}{r} 30+12 \\ -10+\;8 \\ \hline 20+\;4 \end{array}$$

Bringing 1 ten over to the units place, we now have 3 tens left in the tens place. We cross out the 4, write a 3 in the tens place, and write a 1 beside the 2 in the units place. With the ten we "borrowed," we now have 12 in the units place.

$$\begin{array}{r} {}^3\!\!\not{4}\;{}^1\!\!2 \\ -\;1\;8 \\ \hline 2\;4 \\ 4\;2 \end{array} \qquad \begin{array}{r} 30+12 \\ -10+\;8 \\ \hline 20+\;4 \end{array}$$

Now we have enough to count up from 8 to 12. Take the 8 unit bar and the one 10 bar and place them upside down on the 12 units and the 3 remaining tens. The difference is 2 tens and 4 units. Check by adding.

Example 2

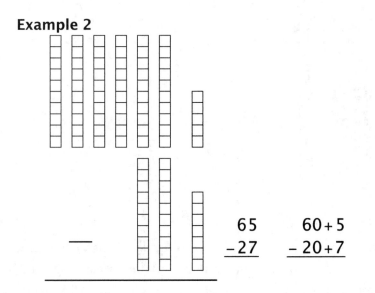

$$\begin{array}{r} 65 \\ -27 \\ \hline \end{array} \qquad \begin{array}{r} 60+5 \\ -20+7 \\ \hline \end{array}$$

The question is, "Can we count 7 from 5?" No, five isn't enough to have 7 counted from it. The number on top has to be the same as or greater than the one beneath it.

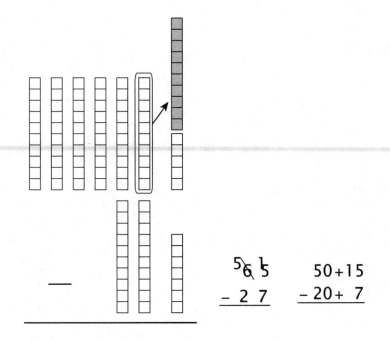

$$\begin{array}{r} {}^5\!\!\!\!6\,{}^1\!5 \\ -\ 2\ 7 \\ \hline \end{array} \qquad \begin{array}{r} 50+15 \\ -20+\ 7 \\ \hline \end{array}$$

Bringing the ten over to the units place, we now have 5 tens left in the tens place. We cross out the 6 , write a 5 in the tens place, and write a 1 beside the 5 in the units place. With the ten we regrouped, we now have 15 in the units place.

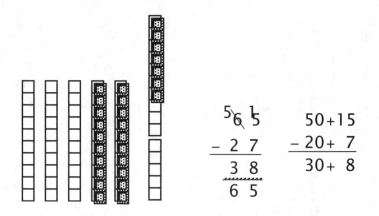

Now we have enough to count up from 7 to 15. Take the 7 unit bar and the two 10 bars and place them upside down on the 15 units and the 5 remaining tens. The difference is 3 tens and 8 units. Check by adding.

Telling Time: Hours

When the minutes are mastered, we can add the hours. Begin by placing a green unit bar at the end of the first five bar (outside the circle), pointing away from the center of the clock. To distinguish between the minutes and hours, I leave the minute unit bars right side up, but I place the hours upside down with the hollow side showing. This distinguishes the minutes and the hours. Place your orange bar (upside down with the hollow side showing) at the end of the second five bar. Continue this process with all the unit bars through 12. See the illustration on this page.

Choose a unit bar smaller than your minute hand for an hour hand. Turn it upside down so the student makes the connection between the hour hand and the hours, since both are upside down.

Position the hour hand so that it points between the 2 and the 3. This is the critical point for telling time. Is it 2 or is it 3? I've explained this to many children, with success, by aiming the hour hand at the 2 and saying, "He just had his second birthday and is now 2." Then I move the hand towards the 3 a little and ask, "How old is he now? Is he still 2?" "Yes." Then I move it a little further and ask if he's still 2. "Yes." I do this until it is almost pointing to 3 and ask the question, "What about the day before his next birthday; how old is he?" "Still 2." He is almost 3 but still 2. Practice this skill by moving the hour hand around the clock until the student can confidently identify the hour.

When the hour hand is mastered, put the hours and minutes together. Have the student identify the hour first and then skip count to find the minutes. Write the time with a colon between the hour and the minutes. For example, two-thirty is written as 2:30. Point out to the student that this is the way time is shown on a digital clock. Use the worksheets for practice. Finally, introduce telling time on a real clock.

There is a removable clock template at the end of the student workbook for you to use, if you wish.

Using A.M. and P.M.

Besides hours and minutes, digital clocks usually show the letters "a.m." or "p.m." Although most clocks measure 12 hours, there are 24 hours between midnight of one day and midnight of the next day. In order to prevent confusion, people began many years ago to use the Latin words "ante meridiem" and "post meridiem." Ante meridiem (a.m.) means "before noon." Times that are noted as a.m. start right after midnight and go until noon. Post meridiem (p.m.)means "after noon." Times that are noted as p.m. start right after noon and go until midnight. After a student can confidently tell time on an analog clock, discuss whether the time is a.m. or p.m. Ask questions about whether a particular activity is more likely to happen in the a.m. or p.m. For example, breakfast is more likely to be at 8:00 a.m., while bedtime is more likely to be 8:00 p.m.

Although noon is, strictly speaking, neither "before noon" nor "after noon," you may see noon indicated by 12:00 p.m. on a digital clock. Similarly, midnight may be given as 12:00 a.m. To prevent confusion, it is better to say "twelve noon" or "twelve midnight."

Subtraction: Three-Digit Numbers
Mental Math

When subtracting two three-digit numbers, we may have to regroup more than once in the same problem. (Example 1 is continued on the next page.)

Example 1

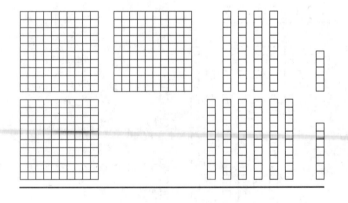

$$
\begin{array}{r}
245 \\
-167 \\
\hline
\end{array}
\qquad
\begin{array}{r}
200+40+5 \\
-100+60+7 \\
\hline
\end{array}
$$

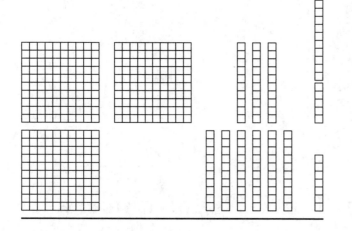

Can we count 7 from 5? No, so we take 1 ten and add it to the 5 units to make 15 units. There are only 3 tens now in the minuend.

$$2\,{}^3\!4\,\cancel{5} \qquad 200+30+15$$
$$-1\,6\,7 \qquad -100+60+\ 7$$

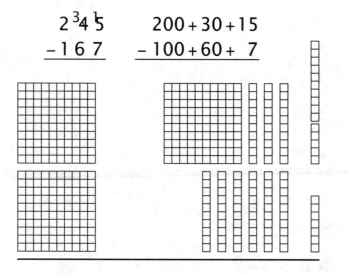

We can't subtract 6 tens from 3 tens. We take 1 hundred and add it to the 3 tens to make 130, or 13 tens. There is only 1 hundred now in the minuend. After this, we subtract. The addition check is also shown.

$$\cancel{2}\,{}^1\!{}^3\!4\,\cancel{5} \qquad 100+130+15$$
$$-1\,6\,7 \qquad -100+\ 60+\ 7$$
$$\overline{7\,8} \qquad \overline{70+\ 8}$$
$$\overline{2\,4\,5}$$

Sometimes we must go all the way to the hundreds place to get what we need for the units place. Study Example 2 carefully to see how to regroup the problem step by step.

Example 2 (continued on the next page)

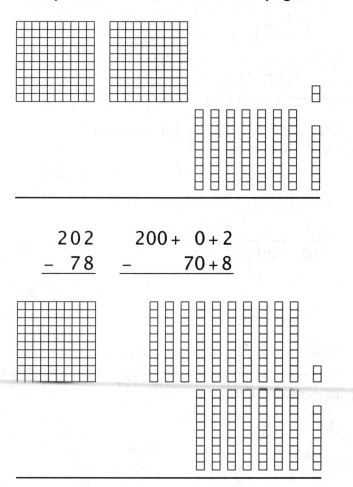

$$\begin{array}{r} 2\,0\,2 \\ -\ 7\,8 \end{array} \qquad \begin{array}{r} 200+\ 0+2 \\ -\quad\ \ 70+8 \end{array}$$

We can't subtract 8 units from 2 units, and there are no tens to regroup and make units. We must take 1 hundred and regroup to make 10 tens.

$$\begin{array}{r} {}^{1}\!\!\not{2}\,\not{0}\,2 \\ -\ 7\,8 \end{array} \qquad \begin{array}{r} 100+100+2 \\ -\quad\ \ 70+8 \end{array}$$

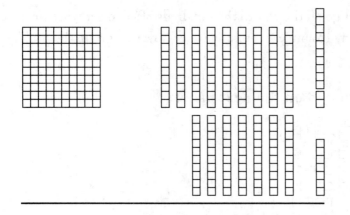

Now we take 1 ten and add it to the 2 units to make 12 units. After this, we subtract.

$$
\begin{array}{r}
9 \\
\overset{\scriptstyle 1}{2}\,\overset{\scriptstyle 1}{0}\,2 \\
-\quad 7\;8 \\
\hline
1\;2\;4 \\
\hline
2\;0\;2
\end{array}
\qquad
\begin{array}{r}
100+90+12 \\
-\qquad 70+\;8 \\
\hline
100+20+\;4
\end{array}
$$

Mental Math

These problems involve mentally adding one to the tens place to find the answer. You many want to start by writing each problem and solving it with the blocks and then move on to doing the problems orally.

1. $120 + 10 = ?$ (130)

2. $236 + 10 = ?$ (246)

3. $350 + 10 = ?$ (360)

4. $146 + 10 = ?$ (156)

5. $453 + 10 = ?$ (463)

6. $705 + 10 = ?$ (715)

7. $532 + 10 = ?$ (542)

8. $788 + 10 = ?$ (798)

Ordinal Numbers; Tally Marks
Days and Months

So far we have been using cardinal numbers, which tell how many. *Ordinal numbers* are used to tell the order of things. Notice the "ord" in ordinal and order. Other than first, second, and third, the most commonly-used names for ordinal numbers end in *th*. I like to use this opportunity to teach days of the week and months of the year. The first day of the week is Sunday, the first month of the year is January, etc. If your student doesn't know the days and months, take some time to learn them before moving to the next lesson. Each day, ask the student to give the date and then use ordinal numbers to tell what month and what day it is. Example: Christmas is in the twelfth month and on the twenty-fifth day.

Choose a birthday or a date in history and ask questions. For example, if a student's birthday is on the eighth month and the first day, what is the date? The answer is August 1.

first	Sunday	January	31
second	Monday	February	28 or 29
third	Tuesday	March	31
fourth	Wednesday	April	30
fifth	Thursday	May	31
sixth	Friday	June	30
seventh	Saturday	July	31
eighth		August	31
ninth		September	30
tenth		October	31
eleventh		November	30
twelfth		December	31

Days in a Month

My grandfather taught me a clever way to tell how many days are in a month without using the poem "30 days has September . . ." Hold up your two fists and, beginning on the left, recite the months. The knuckles, or mountains, have 31 days, and the valleys between the knuckles have 30 days. The exception is February, which has 28 days, or 29 days in a leap year. Notice that when you put your fists together, July and August are both mountains. There is no valley between them, so they each have 31 days.

Example 1
How many days are in September?

Beginning from the left and reciting the months, we end up in a valley, so September has 30 days.

Example 2
What is the tenth month?

Beginning with January as the first month, we end up with October.

Example 3
What is the fourth day of the week?

Beginning with Sunday as the first day of the week, we end up with Wednesday.

Look for other well-known lists to use in order to practice this skill. One example would be the list of U. S. presidents.

Tally Marks

When we use tally marks, we count to four using one line for one, two lines for two, three lines for three, and four lines for four. To show five, we put a slash diagonally through four lines. Tally marks are very useful when keeping track of slowly-changing information. Since cowboys used tallies to keep track of their cattle, I think of cows slowly moving through a gate and a cowboy sitting on a fence, making one mark for each cow passing beneath him. Alternately, perhaps you are on a trip, and you want to count red cars that you pass. Each time you pass a red car, you make a line until you get to the fifth one, and then you make a slash.

Look at the chart below to see how to represent the numbers 1 to 20. Study the examples as we change from tally marks to a numeral and then from a numeral to tally marks.

1	|	11	𝍸 𝍸 |
2	||	12	𝍸 𝍸 ||
3	|||	13	𝍸 𝍸 |||
4	||||	14	𝍸 𝍸 ||||
5	𝍸	15	𝍸 𝍸 𝍸
6	𝍸 |	16	𝍸 𝍸 𝍸 |
7	𝍸 ||	17	𝍸 𝍸 𝍸 ||
8	𝍸 |||	18	𝍸 𝍸 𝍸 |||
9	𝍸 ||||	19	𝍸 𝍸 𝍸 ||||
10	𝍸 𝍸	20	𝍸 𝍸 𝍸 𝍸

Example 1
Show the number 7 using tally marks.

7 = 5 + 2 = 𝍸 ||

Example 2
Write 𝍸 𝍸 |||| as a numeral.

𝍸 𝍸 |||| = 5 + 5 + 4 = 14

Subtraction: Four-Digit Numbers

When we regroup from the thousands to the hundreds, we cross out the digit in the thousands place and replace it with a digit that is one less. Next, we move the thousand we just regrouped to the hundreds place, where it becomes 10 hundreds. Put a 1 beside the digit in the hundreds place to show that it has been increased by 10. This is what we have always done when regrouping tens or hundreds. It shows the uniformity of the base 10 system.

Example 1
Solve: 4,581 – 1,397

$$
\begin{array}{r}
4,5\,\cancel{8}\,\overset{7}{\cancel{1}} \\
-1,3\,9\,7 \\
\hline
\end{array}
\qquad
\begin{array}{r}
4,000+500+70+11 \\
1,000\mid 300\mid 90\mid 7 \\
\hline
\end{array}
$$

We can't subtract 7 units from 1 unit. We take 1 ten and regroup to make 10 units.

$$
\begin{array}{r}
\overset{4\ 17}{4,\cancel{5}\,\cancel{8}\,\overset{7}{\cancel{1}}} \\
-1,3\,9\,7 \\
\hline
\end{array}
\qquad
\begin{array}{r}
4,000+400+170+11 \\
-\ 1,000+300+\ 90+\ 7 \\
\hline
\end{array}
$$

We can't subtract 9 tens from 7 tens. We take 1 hundred and regroup to make 10 more tens.

```
     4 17
   4,5 8 1        4,000+400+170+11
  -1,3 9 7       - 1,000+300+ 90+ 7
   3,1 8 4        3,000+100+ 80+ 4
  ------------
   4,5 8 1
```

The hundreds and thousands are okay, so we can subtract.
We could have subtracted the units after they were changed in
step 1. Either way will work.

Example 2
Solve: 3,062 – 1,549

```
        5
   3,0 6 2        3,000 + 000 + 50 +12
  -1,5 4 9       - 1,000 + 500 + 40 + 9
```

We can't take 9 units from 2 units. We take 1 ten, regroup, and
make 12 units.

```
   2    5
   3, 0 6 2        2,000 +1,000 + 50 +12
  -1, 5 4 9       - 1,000 +  500 + 40 + 9
```

We can subtract 4 tens from 5 tens, so that is fine. However,
we can't take 5 hundreds from 0 hundreds. We regroup 1
thousand to make 10 hundreds.

```
   2    5
   3, 0 6 2        2,000 +1,000 + 50 +12
  -1, 5 4 9       - 1,000 +  500 + 40 + 9
   1, 5 1 3        1,000 +  500 + 10 + 3
  -------------
   3, 0 6 2
```

The thousands are okay, so we can subtract.

LESSON 27

Subtraction: Money
Mental Math

The red hundreds square represents one dollar, the blue tens bar represents one dime, and the green units cube represents one penny. Subtracting money is the same as subtracting three-digit numbers, except for the decimal point. We use the same blocks, but instead of regrouping a hundred to make 10 tens, we regroup one dollar to make 10 dimes. In the same way, we regroup one dime to make 10 pennies, or 10 cents.

Example 1 (continued on the next page)

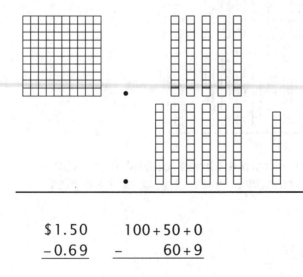

$$
\begin{array}{rr}
\$1.50 & 100+50+0 \\
-\,0.69 & -\quad\;\;60+9 \\
\hline
\end{array}
$$

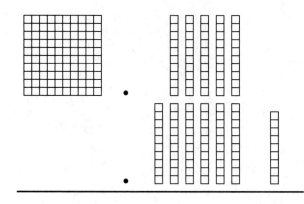

We can't subtract 9¢ from 0¢. We take 1 dime and add it to 0¢ to make 10¢.

```
        4
  $1.5 ⁶0        100+40+10
 - 0.6 9       –     60+ 9
```

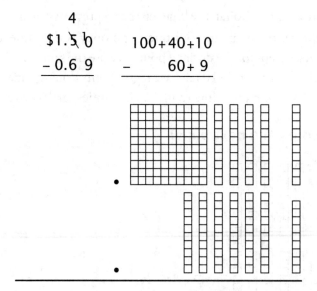

We can't subtract 6 dimes from 4 dimes, or 60¢ from 40¢. We take 1 dollar and add it to 40¢ to make 140¢, or 14 dimes.

```
       14
  $⅟.5 ¹0        140+10
 - 0.6  9       –  60+ 9
  $0. 8 1          80+ 1
```

Mental Math

These problems involve mentally adding one to the hundreds place. A student who understands place value should be able to do these with minimal difficulty.

1. $300 + 100 = ?$ (400)

2. $120 + 100 = ?$ (220)

3. $756 + 100 = ?$ (856)

4. $689 + 100 = ?$ (789)

5. $536 + 100 = ?$ (636)

6. $213 + 100 = ?$ (313)

7. $394 + 100 = ?$ (494)

8. $407 + 100 = ?$ (507)

Subtraction: Multiple-Digit Numbers

The only new part of this lesson is subtracting five digits instead of four. We use the same procedure that we have already learned for regrouping. This time we cross out the number in the ten thousands place and decrease it by one. Put a 1 beside the number in the thousands place to increase it by 10. Everything else is the same as what you have already learned.

Example 1
Solve: 34,085 – 19,760

$$
\begin{array}{r}
3 \\
3\,4,\,085 \\
-\,19,\,760 \\
\end{array}
\qquad
\begin{array}{r}
30,000+3,000+1,000+80+5 \\
-\,10,000+9,000+\;\;700+60+0 \\
\end{array}
$$

The units and tens are okay. We can't subtract 7 hundreds from 0 hundreds, so we take 1 thousand and regroup to make 10 hundreds.

$$
\begin{array}{r}
2\;\;13 \\
3\,4,\,085 \\
-\;\;19,\,760 \\
\hline
14,\,325 \\
\hline
3\,4,\,085 \\
\end{array}
\qquad
\begin{array}{r}
20,000+13,000+1,000+80+5 \\
-\,10,000+\;\;9,000+\;\;700+60+0 \\
\hline
10,000+\;\;4,000+\;\;300+20+5 \\
\end{array}
$$

We can't subtract 9 thousand from 3 thousand, so we take 1 ten thousand and regroup to make 10 thousands. Subtract and then check by adding.

Example 2
Solve: 97,518 – 69,227

```
      4
  97,5̸¹18        90,000+7,000+400+110+8
 -69,2 27       -60,000+9,000+200+ 20+7
 ─────────      ──────────────────────────
```

The units are okay. We can't take 2 tens from 1 ten, so we take
1 hundred and regroup to make 10 more tens.

```
  8   4
  9̸ ⁷,5̸ ¹18      80,000+17,000+400+110+8
 -6 9, 2 27     -60,000+ 9,000+200+ 20+7
 ──────────     ──────────────────────────
  2 8, 2 9 1     20,000+ 8,000+200+ 90+ 1
  ~~~~~~~~~~
  9 7, 5 18
```

We can't subtract 9 thousand from 7 thousand, so we take
1 ten thousand and regroup to make 10 more thousands.
Subtract and then check by adding.

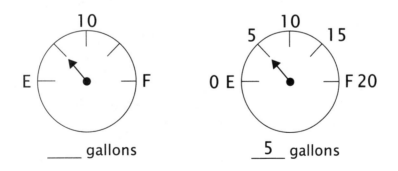

LESSON 29

Reading Gauges and Thermometers

Many measuring instruments do not give you all the information needed. Instead, they require you to figure out some of it. Here are some gauges found around my home and in my car. The key to reading scales on a gauge is finding out what each space stands for on the scale. Skip counting is useful in figuring out the pattern. Once you have figured out the pattern, start at 0, or E, and count how many spaces to the arrow. Be careful that you don't count the lines. Instead count how many spaces are between the lines. For each example, there are two problems and the solutions. Read where the arrow is pointing to find the answer.

Example 1
Here is the 20-gallon gas gauge in my car. According to the gauge, how much gas do I have left?

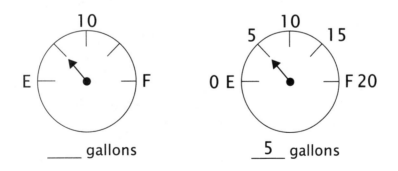

____ gallons   _5_ gallons

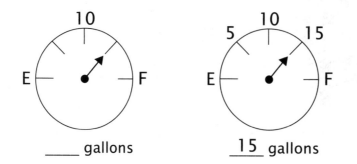

_____ gallons   _15_ gallons

Example 2
The wood stove in the basement has a temperature gauge. What temperature does each gauge show?

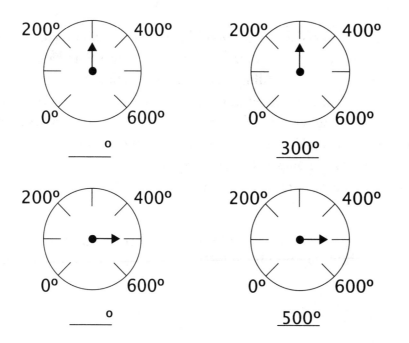

_____ °   _300°_

_____ °   _500°_

Example 3
The speedometer in my car measures speed in miles per hour (mph). What speed does each gauge show?

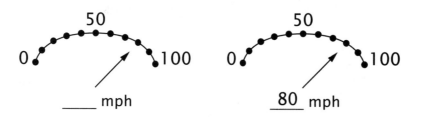

_____ mph   _80_ mph

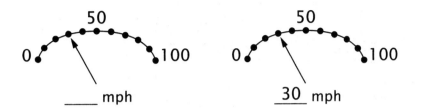

____ mph   30 mph

Thermometers

A thermometer is a special gauge that measures temperature. The unit for temperature is degrees, which is denoted by a floating circle above and to the right of the number.

In Example 4 we recognize skip counting by two: 2, 4, 6, __, __, 12, __, 16. Filling in the missing numbers (0, 2, 4, 6, 8, 10, 12, 14, 16), we discover that the temperature is 10°.

Example 4

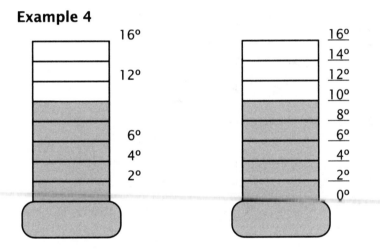

The thermometer reads 10°.

Example 5

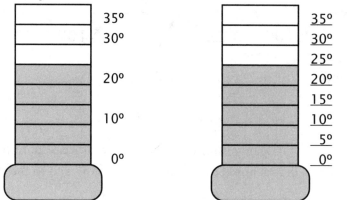

The thermometer reads 25°.

Example 6

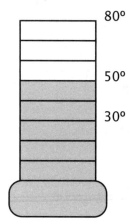

The thermometer reads 50°.

Bar Graphs and Line Graphs

A bar graph is used to compare number values. When making a bar graph, use your bars to represent the values. To make this lesson real and practical, choose from the following list of ideas or make up some of your own. Compare the amount of money earned in a week, rainfall, how fast you grow, temperature, home runs of your favorite baseball team, or the number correct on your assignments. Make it REAL. If I were to compare the ages of four boys, the graph might look like this. Notice that a graph always has a title.

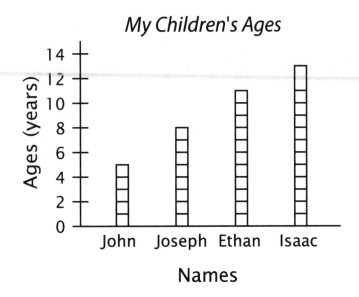

The objects, or data, are on the bottom and labeled. The scale is on the left. For greater numbers, let each block or unit represent 5, 10, or 100.

Line Graphs

A line graph is used to show change over time. This graph shows how much a tomato plant grew during the month of June.

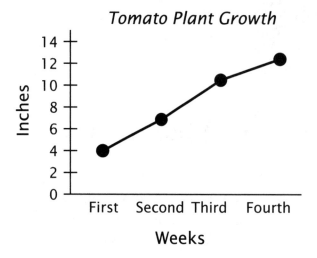

As you record your own data, you should choose bar graphs for information that compares separate pieces of data, such as the amount of money different people earn in a week. A line graph is best for data that changes gradually over time, such one person's change in height.

Information may also be recorded using pictures or dots to indicate how many times a certain number occurs. These kinds of graphs are called picture graphs and dot plots. Look for graphs in newspapers and other textbooks your students might be using.

Identify Shapes; Fractional Parts

Students have been taught to identify both squares and rectangles as shapes with four straight sides and four square corners. A square is a special kind of rectangle where all four sides have the same length.

A more general name for any shape with four straight sides is *quadrilateral*. The sides do not have to have any particular lengths, and the corners may not be square corners (right angles). Here are some examples of quadrilaterals.

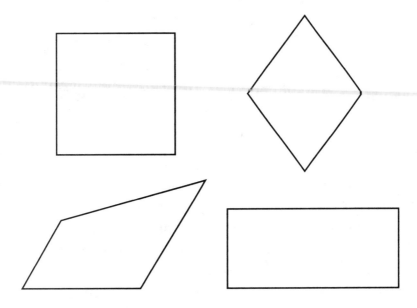

Notice that squares and rectangles are quadrilaterals, but not all quadrilaterals are squares or rectangles. The key defining characteristic is the number of sides.

Here are some other shapes that may be named by the numbers of their sides.

triangle - three sides

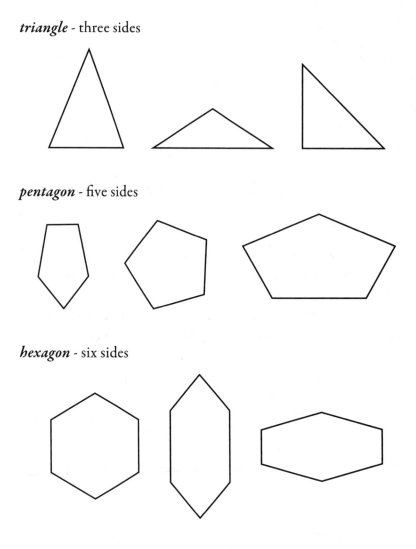

pentagon - five sides

hexagon - six sides

Stress that it is the number of sides that is important in naming the shapes. If a pentagon has five equal sides, it is a regular pentagon. However, it is still a pentagon if the sides are not equal. Attribute blocks and other resources that students may use at this level often present regular shapes. Point out that shapes with the same number of sides may look quite different from each other.

All of the shapes mentioned so far are two-dimensional. At this level, a student should also be able to identify a *cube*. A cube is a simple three-dimensional shape made up of six faces that are all squares. All of the twelve edges of a cube are the same length. The dice used for games are good examples of cubes.

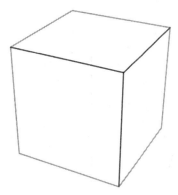

You can cut out different two-dimensional shapes from construction paper and have students use them to practice identification. Commercially-available pattern blocks or attribute blocks are a useful and fun way to become familiar with the various shapes. Many of these resources are also available to download from the internet.

Fractional Parts - Sharing Shapes

Math-U-See does not teach operations with fractions until *Epsilon*. However, students should become familiar with commonly-used words that indicate equal parts of a circle or rectangle. Introduce or review *half of*, *halves*, *a third of*, *thirds*, *a fourth of (a quarter of)*, *fourths (quarters)*.

The idea of fractional parts is easily connected to the idea of sharing equally or fairly, a concept that is very dear to the heart of children. Be sure to use the terms above when cutting a pizza or a pan of brownies. On worksheets, ask students to color a particular fractional part of a circle or rectangle.

Example 1
Color one third of the rectangle.

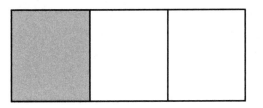

Example 2
Color one fourth of the circle.

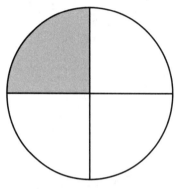

Be clear that one half means one of two equal shares, not necessarily two pieces with a certain shape. Study the example.

Example 3
A rectangular cake is cut into 12 pieces. Divide the pieces into two equal shares. Here are two possible answers.

Each drawing shows two equal shares with six pieces of cake in each share. The shape of the share is different in each drawing.

Number Line

Young children usually begin their understanding of numbers with concrete objects such as apples or pencils. They must then bridge the gap from the concrete to the abstract representation of numbers with symbols. The Math-U-See blocks are an important part of this process, as they are concrete objects that also clearly represent specific numbers. The number line is another tool that can help students move from the concrete to the abstract.

Number lines will be important for concepts that are taught in future levels of Math-U-See. It is also interesting to note that the thermometers in lessons 29 and 30 in the student workbook are vertical number lines. The numbers start with zero and move up the scale, rather than going left to right. Later on, when students are introduced to negative numbers, the "below zero" temperatures are a useful real-life example. The scales for the graphs in lesson 30 are also examples of number lines. Notice that some are horizontal and some are vertical.

A basic number line is similar to a ruler. It is numbered in increments starting with zero at the left and counting to the right as far as is convenient. To introduce a number line, start with the blue ten bar. Point out that it represents the number 10. Use the ten bar to draw a number line the same length as the bar. Mark off 10 spaces and number them as shown below. We use an arrow on the end of the line to show that we could keep on marking and numbering spaces as long as we wished.

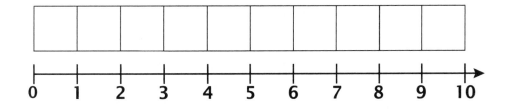

To show addition on the number line, begin with the blocks and remind students of how they are used to show addition. Here are three ways to show six plus four: with the blocks, on a number line, and abstractly with symbols.

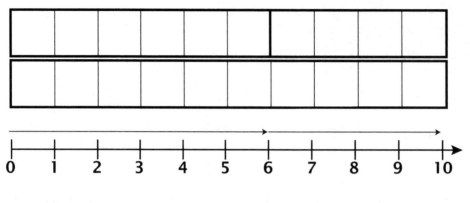

6 + 4 = 10

A number line also clearly shows the difference between two numbers. Here is the difference between five and two shown with the blocks, on the number line, and with numerals. To show subtraction with the blocks, turn the two bar upside down so that the holes are on top and put it on top of the five bar.

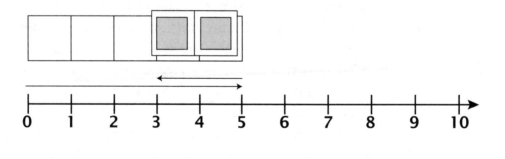

5 − 2 = 3

Student Solutions

Lesson Practice 1A

1. 146
   "one hundred forty-six"
2. 254
   "two hundred fifty-four"
3. 2 tens and 8 units
   "twenty-eight"
4. 3 hundreds, 3 tens, and 6 units
   "three hundred thirty-six"
5. $3 + 1 = 4$
6. $5 + 0 = 5$
7. $2 + 8 = 10$
8. $0 + 7 = 7$
9. $5 + 2 = 7$
10. $2 + 1 = 3$
11. $1 + 5 = 6$
12. $6 + 2 = 8$
13. $4 + 2 = 6$
14. $2 + 9 = 11$
15. $7 + 2 = 9$

Lesson Practice 1B

1. 139
   "one hundred thirty-nine"
2. 325
   "three hundred twenty-five"
3. 3 tens and 1 unit
   "thirty-one"
4. 2 hundreds, 2 tens, and 8 units
   "two hundred twenty-eight"
5. $9 + 3 = 12$
6. $8 + 7 = 15$
7. $9 + 5 = 14$
8. $8 + 3 = 11$
9. $5 + 8 = 13$
10. $6 + 9 = 15$
11. $9 + 4 = 13$
12. $8 + 9 = 17$
13. $9 + 7 = 16$
14. $6 + 8 = 14$
15. $8 + 4 = 12$

Lesson Practice 1C

1. 193
   "one hundred ninety-three"
2. 262
   "two hundred sixty-two"
3. 4 tens and 9 units
   "forty-nine"
4. 2 hundreds, 8 tens, and 3 units
   "two hundred eighty-three"
5. $8 + 8 = 16$
6. $7 + 7 = 14$
7. $7 + 6 = 13$
8. $4 + 4 = 8$
9. $5 + 5 = 10$
10. $6 + 5 = 11$
11. $3 + 3 = 6$
12. $2 + 2 = 4$
13. $6 + 6 = 12$
14. $3 + 4 = 7$
15. $9 + 9 = 18$

Lesson Practice 1D

1. 105
   "one hundred five"
2. 246
   "two hundred forty-six"
3. 7 tens and 2 units
   "seventy-two"
4. 4 hundreds, 1 ten, and 7 units
   "four hundred seventeen"
5. $8 + 1 = 9$
6. $7 + 3 = 10$
7. $3 + 6 = 9$
8. $1 + 9 = 10$
9. $6 + 4 = 10$
10. $5 + 5 = 10$
11. $7 + 2 = 9$
12. $5 + 4 = 9$
13. $2 + 8 = 10$
14. $3 + 7 = 10$
15. $4 + 6 = 10$

Systematic Review 1E

1. 318
   "three hundred eighteen"
2. 273
   "two hundred seventy-three"
3. 5 tens; "fifty"
4. 1 hundred and 9 tens
   "one hundred ninety"
5. $4 + 7 = 11$
6. $7 + 5 = 12$
7. $3 + 5 = 8$
8. $5 + 7 = 12$
9. $5 + 3 = 8$
10. $7 + 4 = 11$
11. $9 + 8 = 17$
12. $5 + 3 = 8$
13. $7 + 6 = 13$
14. $8 + 2 = 10$
15. $1 + 0 = 1$

Systematic Review 1F

1. 130
   "one hundred thirty"
2. 45; "forty-five"
3. 3 hundreds and 6 units
   "three hundred six"
4. 2 hundreds, 2 tens, and 2 units
   "two hundred twenty-two"
5. $1 + 1 = 2$
6. $7 + 2 = 9$
7. $9 + 5 = 14$
8. $3 + 8 = 11$
9. $6 + 6 = 12$
10. $7 + 8 = 15$
11. $5 + 4 = 9$
12. $3 + 5 = 8$
13. $5 + 8 = 13$
14. $9 + 4 = 13$
15. $3 + 3 = 6$
16. $6 + 7 = 13$
17. $7 + 5 = 12$
18. $4 + 8 = 12$

Lesson Practice 2A

1. 4, 12, 20
2. 6, 31, 206
3. 162, 55, 4
4. 136, 16, 3
5. 8, 9, 10, 11, 12
6. 10, 9, 8, 7, 6
7. 71, 72, 73, 74, 75

Lesson Practice 2B

1. 5, 8, 16
2. 25, 63, 100
3. 7, 17 107
4. 11, 89, 200
5. 30, 10, 6
6. 80, 20, 10
7. 245, 61, 9
8. 100, 84, 3
9. 6, 5, 4, 3, 2
10. 22, 23, 24, 25, 26
11. 64, 65, 66, 67, 68
12. 17, 16, 15, 14, 13

Lesson Practice 2C

1. 6, 50, 100
2. 18, 32, 95
3. 10, 40, 400
4. 16, 65, 243
5. 125, 15, 5
6. 170, 106, 9
7. 99, 48, 22
8. 120, 114, 89
9. 29, 28, 27, 26, 25
10. 43, 44, 45, 46, 47
11. 30, 31, 32, 33, 34
12. 9, 8, 7, 6, 5

Systematic Review 2D

1. 5, 15, 17
2. 8, 11, 40
3. 75, 43, 12
4. 200, 150, 110
5. 79, 80, 81, 82, 83
6. 4 tens and 7 units
   "forty-seven"
7. 1 hundred, and 9 units
   "one hundred nine"
8. $9 + 9 = 18$
9. $7 + 8 = 15$
10. $4 + 5 = 9$
11. $2 + 3 = 5$
12. $4 + 8 = 12$
13. $5 + 7 = 12$
14. $3 + 8 = 11$
15. 121 is greater than 112; Sam
16. $4 + 1 = 5$ sandwiches

Systematic Review 2E

1. 5, 18, 63
2. 100, 200, 400
3. 56, 44, 14
4. 200, 105, 50
5. 9, 8, 7, 6, 5
6. 9 tens and 8 units
   "ninety-eight"
7. 2 hundreds, 7 tens, and 6 units
   "two hundred seventy-six"
8. $6 + 6 = 12$
9. $5 + 9 = 14$
10. $2 + 7 = 9$
11. $4 + 6 = 10$
12. $8 + 9 = 17$
13. $5 + 6 = 11$
14. $1 + 7 = 8$
15. 7 is less than 16; daisies
16. $7 + 9 = 16$ vehicles

Systematic Review 2F

1. 19, 39, 99
2. 10, 60, 80
3. 299, 74, 18
4. 180, 48, 21
5. 88, 89, 90, 91, 92
6. 2 hundreds and 4 tens
   "two hundred forty"
7. 1 ten and 6 units
   "sixteen"
8. $7 + 7 = 14$
9. $2 + 5 = 7$
10. $3 + 7 = 10$
11. $0 + 4 = 4$
12. $8 + 8 = 16$
13. $6 + 7 = 13$
14. $4 + 9 = 13$
15. 62 is greater than 26; story books
16. $6 + 8 = 14$ coins

Lesson Practice 3A

1. $4 > 3$
2. $7 = 7$
3. $6 > 4$
4. $11 < 21$
5. $14 = 14$
6. $10 > 9$
7. $63 > 36$
8. $12 = 12$
9. $6 < 7$
10. $12 > 10$
11. $50¢ < 90¢$
    $90¢ > 50¢$
12. Denise: $5 + 4 = 9$
    Michael: $5 + 5 = 10$
    $9 < 10$; Michael did more.
13. $12 > 8$; Ethan read fewer.
14. $14 = 14$

Lesson Practice 3B

1. $5 < 7$
2. $2 < 4$
3. $10 > 9$
4. $13 > 10$
5. $18 > 17$
6. $11 = 11$
7. $27 < 72$
8. $14 < 15$
9. $6 = 6$
10. $12 = 12$
11. William: $2 + 1 = 3$
    Brother: $3 + 2 = 5$
    $5 > 3$; brother had more.
12. $9 > 8$; Alexa got more right.
13. $10 = 10$
14. $150 > 105$

Lesson Practice 3C

1. $9 < 14$
2. $5 < 6$
3. $9 > 7$
4. $25 > 15$
5. $16 = 16$
6. $15 < 17$
7. $31 > 13$
8. $12 = 12$
9. $7 > 5$
10. $11 < 12$
11. $10 < 110$; Rachel has fewer.
12. $65 > 55$; more on clothes
13. $10 = 10$
14. $213 < 231$

Systematic Review 3D

1. $13 > 12$
2. $57 < 75$
3. 7, 8, 13
4. 50, 70, 90
5. 28, 29, 30, 31, 32
6. 2 hundreds, 1 ten, and 1 unit
   "two hundred eleven"

7. 3 tens and 6 units
   "thirty-six"
8. $2 + 9 = 11$
9. $5 + 5 = 10$
10. $3 + 6 = 9$
11. $2 + 8 = 10$
12. $6 + 9 = 15$
13. $3 + 4 = 7$
14. $0 + 6 = 6$
15. $452 > 254$
16. $5 + 3 = 8$ children

Systematic Review 3E

1. $8 = 8$
2. $91 > 19$
3. 200, 12, 2
4. 105, 51, 15
5. 17, 18, 19, 20, 21
6. 1 hundred and 4 units
   "one hundred four"
7. 1 ten and 3 units
   "thirteen"
8. $4 + 4 = 8$
9. $1 + 6 = 7$
10. $2 + 4 = 6$
11. $0 + 1 = 1$
12. $9 + 9 = 18$
13. $7 + 8 = 15$
14. $4 + 7 = 11$
15. $3 > 2$; more brothers
16. $3 + 3 = 6$ miles

Systematic Review 3F

1. $6 < 7$
2. $15 = 15$
3. 300, 63, 3
4. 197, 91, 74
5. 13, 12, 11, 10, 9
6. 1 hundred, 3 tens, and 5 units
   "one hundred thirty-five"
7. 2 hundreds and 4 tens
   "two hundred forty"

8. $7 + 9 = 16$
9. $2 + 2 = 4$
10. $4 + 9 = 13$
11. $3 + 7 = 10$
12. $1 + 1 = 2$
13. $5 + 8 = 13$
14. $7 + 7 = 14$
15. $13 = 13$
16. $6 + 8 = 14$ pages

Lesson Practice 4A

1. 40
2. 40
3. done
4. 60
5. done
6.
$$\begin{array}{r} (10) \\ +(10) \\ \hline (20) \end{array}$$
7.
$$\begin{array}{r} (30) \\ +(20) \\ \hline (50) \end{array}$$
8.
$$\begin{array}{r} (50) \\ +(40) \\ \hline (90) \end{array}$$
9.

3	5	8
6	1	7
9	6	15

10. $(20) + (30) = (50)$ books

Lesson Practice 4B

1. 40
2. 20
3. 60
4. 90
5.
$$\begin{array}{r} (20) \\ +(20) \\ \hline (40) \end{array}$$
6.
$$\begin{array}{r} (50) \\ +(30) \\ \hline (80) \end{array}$$

7.
$$\begin{array}{r} (30) \\ +(20) \\ \hline (50) \end{array}$$
8.
$$\begin{array}{r} (60) \\ +(20) \\ \hline (80) \end{array}$$
9.

3	2	5
5	2	7
8	4	12

10. $(60) + (30) = (90)$ dollars

Lesson Practice 4C

1. 30
2. 80
3. 70
4. 60
5.
$$\begin{array}{r} (30) \\ +(30) \\ \hline (60) \end{array}$$
6.
$$\begin{array}{r} (20) \\ +(10) \\ \hline (30) \end{array}$$
7.
$$\begin{array}{r} (40) \\ +(50) \\ \hline (90) \end{array}$$
8.
$$\begin{array}{r} (20) \\ +(40) \\ \hline (60) \end{array}$$
9.

4	5	9
3	1	4
7	6	13

10. $(30) + (30) = (60)$ in

Systematic Review 4D

1. 80
2. 50
3.
$$\begin{array}{r} (50) \\ +(20) \\ \hline (70) \end{array}$$

4.
$$\begin{array}{r} (20) \\ +(30) \\ \hline (50) \end{array}$$

5.

6	1	7
2	3	5
8	4	12

6. $4 = 4$

7. $16 < 21$

8. $4 + 6 = 10$

9. $1 + 9 = 10$

10. $6 + 9 = 15$

11. $5 + 7 = 12$

12. $(20) + (10) = (30)$ hrs

13. $9 + 8 = 17$ battles

14. $2 + 2 = 4$
$4 + 6 = 10$ birds

Systematic Review 4E

1. 50

2. 60

3.
$$\begin{array}{r} (40) \\ +(10) \\ \hline (50) \end{array}$$

4.
$$\begin{array}{r} (70) \\ +(10) \\ \hline (80) \end{array}$$

5.

2	1	3
7	0	7
9	1	10

6. $14 > 13$

7. $99 < 105$

8. $3 + 9 = 12$

9. $1 + 8 = 9$

10. $5 + 5 = 10$

11. $2 + 7 = 9$

12. $3 + 3 = 6$ cones

13. $(20) + (30) = (50)$ worms

14. $410 > 279$

Systematic Review 4F

1. 90

2. 80

3.
$$\begin{array}{r} (50) \\ +(30) \\ \hline (80) \end{array}$$

4.
$$\begin{array}{r} (50) \\ +(20) \\ \hline (70) \end{array}$$

5.

3	3	6
2	1	3
5	4	9

6. $18 > 16$

7. $198 < 201$

8. $9 + 5 = 14$

9. $2 + 6 = 8$

10. $4 + 3 = 7$

11. $3 + 8 = 11$

12. $9 + 2 = 11$ cards

13. $(50) + (40) = (90)$ radios

14. $45 < 405$

Lesson Practice 5A

1. done

2. $40 + 7 = 47$

3. $100 + 40 + 3$

4. $60 + 5$

5. done

6. $70 + 9 = 79$

7. $40 + 5 = 45$

8. $200 + 60 + 9 = 269$

9. $24 + 33 = 57$

10. $16 + 12 = 28$ dollars

Lesson Practice 5B

1. $100 + 20 + 4 = 124$

2. $50 + 9 = 59$

3. $300 + 60 + 5$

4. $40 + 1$

5. $70 + 9 = 79$

6. $20 + 6 = 26$

7. $60 + 8 = 68$

8. $300 + 50 + 5 = 355$

9. $133 + 10 = 143$ dollars

10. $23 + 23 = 46$ blackbirds

Lesson Practice 5C

1. $300 + 30 + 8 = 338$
2. $60 + 8 = 68$
3. $100 + 70 + 2$
4. $20 + 7$
5. $40 + 4 = 44$
6. $90 + 6 = 96$
7. $90 + 9 = 99$
8. $400 + 40 + 8 = 448$
9. $12 + 36 = 48$ students
10. $41¢ + 46¢ = 87¢$

Systematic Review 5D

1. $500 + 30 + 1$
2. $10 + 8$
3. $40 + 3 = 43$
4. $400 + 20 + 2 = 422$
5. done
6. $(40) + (20) = (60)$
   $43 + 21 = 64$
7. $7 < 12$
8. $71 > 17$
9. $6 + \underline{6} = 12$
10. $7 + \underline{4} = 11$
11. $B + 5 = 7$
   $B = 2$ books
12. $120 + 254 = 374$ birds

Systematic Review 5E

1. $100 + 10 + 4$
2. $30 + 9$
3. $30 + 9 = 39$
4. $300 + 50 + 3 = 353$
5. $(70) + (10) = (80)$
   $68 + 11 = 79$
6. $(50) + (30) = (80)$
   $51 + 34 = 85$
7.

5	6	11
7	1	8
12	7	19

8. $15 = 15$
9. $103 < 331$

10. $7 + \underline{3} = 10$
11. $9 + \underline{7} = 16$
12. $5 + S = 9$
   $S = 4$ spiders
13. $(20) + (20) = (40)$ questions
14. $212 + 362 = 574$ miles

Systematic Review 5F

1. $300 + 10 + 4$
2. $70 + 2$
3. $90 + 9 = 99$
4. $400 + 90 + 9 = 499$
5. $(70) + (20) = (90)$
   $72 + 15 = 87$
6. $(60) + (20) = (80)$
   $61 + 17 = 78$
7.

2	8	10
6	3	9
8	11	19

8. $6 = 6$
9. $86 > 68$
10. $8 + \underline{5} = 13$
11. $4 + \underline{1} = 5$
12. $9 + P = 11$; $P = 2$ peanuts
13. $665 + 133 = 798$ animals
14. $42 > 24$

Lesson Practice 6A

1. 2, 4, 6, 8, 10, 12, 14, 16, 18, 20
2. 2, 4, 6, 8, 10, 12, 14, 16, 18, 20
3. 2, 4, 6, 8, 10, 12, 14, 16, 18, 20
4. 2, 4, 6, 8, $\underline{10}$ cookies
5. 2, 4, 6, 8, 10, $\underline{12}$ X's
6. 2, 4, 6, 8, 10, 12, 14, 16, $\underline{18}$ shoes

Lesson Practice 6B

1. 2, 4, 6, 8, 10, 12, 14, 16, 18, 20
2. 2, 4, 6, 8, 10, 12, 14, 16, 18, 20
3. 2, 4, 6, 8, 10, 12, 14, 16, 18, 20
4. 2, 4, $\underline{6}$ cookies
5. 2, 4, 6, 8, 10, 12, $\underline{14}$ X's
6. 2, 4, 6, 8, $\underline{10}$ ears

Lesson Practice 6C

1. 2, 4, 6, 8, 10, 12, 14, 16, 18, 20
2. 2, 4, 6, 8, 10, 12, 14, 16, 18, 20
3. 2, 4, 6, 8, 10, 12, 14, 16, 18, 20
4. 2, 4, 6, 8, 10, 12, 14, <u>16</u> snowflakes
5. 2, 4, 6, 8, <u>10</u> arms
6. 2, 4, <u>6</u> books

Systematic Review 6D

1. 2, 4, 6, 8, 10, 12, 14, 16, 18, 20
2. $20 + 2 = 22$
3. $200 + 80 + 9 = 289$
4. $(10) + (20) = (30)$
   $13 + 15 = 28$
5. $(30) + (40) = (70)$
   $32 + 41 = 73$
6. $8 > 6$
7. $108 < 801$
8. $5 + \underline{3} = 8$
9. $8 + \underline{7} = 15$
10. $9 + \underline{4} = 13$
11. 2, 4, 6, 8, 10, 12, <u>14</u> times
12. $7 + 2 = 9$
    $9 + 3 = 12$ times

Systematic Review 6E

1. 2, 4, 6, 8, 10, 12, 14, 16, 18, 20
2. $30 + 8 = 38$
3. $600 + 90 + 9 = 699$
4. $(50) + (20) = (70)$
   $48 + 21 = 69$
5. $(50) + (10) = (60)$
   $54 + 12 = 66$
6. $10 > 8$
7. $295 < 592$
8. $3 + \underline{0} = 3$
9. $5 + \underline{3} = 8$
10. $4 + \underline{5} = 9$
11. $126 + 132 = 258$ dollars
12. 2, 4, 6, 8, 10, <u>12</u> fawns

Systematic Review 6F

1. 2, 4, 6, 8, 10, 12, 14, 16, 18, 20
2. $60 + 3 = 63$
3. $600 + 90 + 6 = 696$
4. $(10) + (30) = (40)$
   $11 + 33 = 44$
5. $(50) + (30) = (80)$
   $53 + 25 = 78$
6. $7 = 7$
7. $42 > 24$
8. $7 + \underline{2} = 9$
9. $6 + \underline{6} = 12$
10. $7 + \underline{7} = 14$
11. 2, 4, 6, <u>8</u> inches
12. $11 + 6 = 17$ pennies

Lesson Practice 7A

1. done

2.
$$\begin{array}{cc} \overset{1}{}46 & \overset{10}{}40+6 \\ +35 & +30+5 \\ \hline 81 & 80+1 \end{array}$$

3.
$$\begin{array}{cc} \overset{1}{}73 & \overset{10}{}70+3 \\ +18 & +10+8 \\ \hline 91 & 90+1 \end{array}$$

4.
$$\begin{array}{cc} \overset{1}{}35 & \overset{10}{}30+5 \\ +37 & +30+7 \\ \hline 72 & 70+2 \end{array}$$

5.
$$\begin{array}{cc} \overset{1}{}26 & \overset{10}{}20+6 \\ +55 & +50+5 \\ \hline 81 & 80+1 \end{array}$$

6.
$$\begin{array}{cc} \overset{1}{}38 & \overset{10}{}30+8 \\ +44 & +40+4 \\ \hline 82 & 80+2 \end{array}$$

7.
$$\begin{array}{c} \overset{1}{}47 \\ +25 \\ \hline 72 \end{array}$$

8.
$$\begin{array}{c} 66 \\ +33 \\ \hline 99 \end{array}$$

9.
```
  1
  7 5
+ 1 6
-----
  9 1
```

10. 59 + 37 = 96 marbles

Lesson Practice 7B

1.
```
  1      10
  3 5    30 + 5
+ 2 5  + 20 + 5
-----  --------
  6 0    60 + 0
```

2.
```
  1      10
  5 9    50 + 9
+ 2 4  + 20 + 4
-----  --------
  8 3    80 + 3
```

3.
```
  1      10
  4 5    40 + 5
+ 2 8  + 20 + 8
-----  --------
  7 3    70 + 3
```

4.
```
  1      10
  7 6    70 + 6
+ 1 5  + 10 + 5
-----  --------
  9 1    90 + 1
```

5.
```
  1      10
  2 4    20 + 4
+   9  + 00 + 9
-----  --------
  3 3    30 + 3
```

6.
```
  1      10
  2 5    20 + 5
+ 3 9  + 30 + 9
-----  --------
  6 4    60 + 4
```

7.
```
  1
  2 4
+ 4 8
-----
  7 2
```

8.
```
  1
  6 7
+   8
-----
  7 5
```

9.
```
  1
  5 6
+ 2 4
-----
  8 0
```

10. 26 + 38 = 64 cookies

Lesson Practice 7C

1.
```
  1      10
  2 5    20 + 5
+ 6 5  + 60 + 5
-----  --------
  9 0    90 + 0
```

2.
```
  3 4    30 + 4
+ 4 5  + 40 + 5
-----  --------
  7 9    70 + 9
```

3.
```
  1      10
  7 2    70 + 2
+ 1 8  + 10 + 8
-----  --------
  9 0    90 + 0
```

4.
```
  1      10
  6 8    60 + 8
+   8  + 00 + 8
-----  --------
  7 6    70 + 6
```

5.
```
  1      10
  3 4    30 + 4
+ 4 6  + 40 + 6
-----  --------
  8 0    80 + 0
```

6.
```
  1      10
  3 6    30 + 6
+ 4 5  + 40 + 5
-----  --------
  8 1    80 + 1
```

7.
```
  1
  4 8
+   7
-----
  5 5
```

8.
```
  1
  2 4
+ 6 6
-----
  9 0
```

9.
```
  1
  1 8
+ 5 3
-----
  7 1
```

10. 16 + 8 = 24 cards

Systematic Review 7D

1.
```
  1      10
  4 9    40 + 9
+ 4 5  + 40 + 5
-----  --------
  9 4    90 + 4
```

2.
```
  1      10
  3 6    30 + 6
+ 3 6  + 30 + 6
-----  --------
  7 2    70 + 2
```

3.
```
   1        10
  68      60 + 8
+ 25     + 20 + 5
  93      90 + 3
```

4.
```
  55
+ 14
  69
```

5.
```
   1
  77
+  7
  84
```

6.
```
  95
+  3
  98
```

7. 2, 4, 6, 8, 10, 12, 14, 16, 18, 20

8. 12 < 13

9. 113 > 103

10. 8 + 6 = 14

11. 6 + 4 = 10

12. 9 + 6 = 15

13.
1	4	5
7	3	10
8	7	15

14. 2, 4, 6, 8, 10, 12 jokes

15. 14 + 17 = 31 trips

Systematic Review 7E

1.
```
  12
+ 13
  25
```

2.
```
   1
  37
+ 28
  65
```

3.
```
  63
+ 36
  99
```

4. 22 + 48 = 70

5. 82 + 9 = 91

6. 52 + 6 = 58

7. 2, 4, 6, 8, 10, 12, 14, 16, 18, 20

8. 20

9. 20

10. 9 + 5 = 14

11. 5 + 1 = 6

12. 7 + 6 = 13

13.
2	6	8
5	8	13
7	14	21

14. 11 > 8

15. 125 + 172 = 297 dollars

Systematic Review 7F

1.
```
  43
+ 26
  69
```

2.
```
   1
  19
+ 18
  37
```

3.
```
  62
+ 32
  94
```

4. 15 + 36 = 51

5. 29 + 8 = 37

6. 79 + 6 = 85

7. 2, 4, 6, 8, 10, 12, 14, 16, 18, 20

8. 50

9. 70

10. 7 + 5 = 12

11. 8 + 3 = 11

12. 9 + 8 = 17

13.
7	8	15
6	5	11
13	13	26

14. 2, 4, 6, 8, 10, 12, 14, 16 bows

15. (30) + (10) = (40)

28 + 13 = 41 animals

Lesson Practice 8A

1. 10, 20, 30, 40, 50, 60, 70, 80, 90, 100

2. 10, 20, 30, 40, 50, 60, 70, 80, 90, 100

3. 10, 20, 30, 40, 50, 60, 70¢

4. 10, 20, 30, 40¢

5. 10, 20, 30, 40, 50¢

6. 10, 20, 30; 3 packages

Lesson Practice 8B
1. 10, 20, 30, 40, 50, 60, 70, 80, 90, 100
2. 10, 20, 30, 40, 50, 60, 70, 80, 90, 100
3. 10, 20, 30, 40, 50, <u>60</u>¢
4. 10, 20, <u>30</u>¢
5. 10, 20, 30, 40, 50, 60, 70, 80¢
6. 10, 20, 30, 40, 50, 60, 70,
   80, 90, 100 pennies

Lesson Practice 8C
1. 10, 20, 30, 40, 50, 60, 70, 80, 90, 100
2. 10, 20, 30, 40, 50, 60, 70, 80, 90, 100
3. 10, 20, 30, 40, <u>50</u>¢
4. 10, 20, 30, 40, 50, 60, 70, 80, <u>90</u>¢
5. 10, 20, 30 pennies
6. 10, 20, 30, 40, 50, 60 dollars

Systematic Review 8D
1. 10, 20, 30, 40, 50, 60, 70, 80, 90, 100
2. 2, 4, 6, 8, 10, 12, 14, 16, 18, 20
3.
$$\begin{array}{r} 11 \\ +32 \\ \hline 43 \end{array}$$
4.
$$\begin{array}{r} {}^1 \\ 64 \\ +26 \\ \hline 90 \end{array}$$
5.
$$\begin{array}{r} {}^1 \\ 55 \\ +17 \\ \hline 72 \end{array}$$
6.
$$\begin{array}{r} 341 \\ +111 \\ \hline 452 \end{array}$$
7.
$$\begin{array}{r} 629 \\ +250 \\ \hline 879 \end{array}$$
8.
$$\begin{array}{r} 165 \\ +522 \\ \hline 687 \end{array}$$
9. $6 + \underline{1} = 7$
10. $6 + \underline{2} = 8$
11. $3 + \underline{0} = 3$

12. 10, <u>20</u> pennies
13. $(20) + (40) = (60)$ dollars
    $19 + 36 = 55$ dollars
14. $41 + 48 = 89$ birds
15. 9 dimes = 90¢
    95 pennies = 95¢
    95¢ > 90¢; so 95 pennies

Systematic Review 8E
1. 2, 4, 6, 8, 10, 12, 14, 16, 18, 20
2. 10, 20, 30, 40, 50, 60, 70, 80, 90, 100
3.
$$\begin{array}{r} 15 \\ +64 \\ \hline 79 \end{array}$$
4.
$$\begin{array}{r} {}^1 \\ 13 \\ +28 \\ \hline 41 \end{array}$$
5.
$$\begin{array}{r} {}^1 \\ 44 \\ +46 \\ \hline 90 \end{array}$$
6.
$$\begin{array}{r} 175 \\ +114 \\ \hline 289 \end{array}$$
7.
$$\begin{array}{r} 732 \\ +156 \\ \hline 888 \end{array}$$
8.
$$\begin{array}{r} 244 \\ +234 \\ \hline 478 \end{array}$$
9. $4 + \underline{2} = 6$
10. $6 + \underline{4} = 10$
11. $\underline{9} + \underline{9} = 18$
12. $6 + 2 = 8$
    $8 + 3 = 11$ toys
13. $2 + 5 = 7$ dimes
    10, 20, 30, 40, 50, 60, <u>70</u>¢
14. $8 + \underline{4} = 12$; so 4 gifts
15. $242 + 157 = 399$ trees

Systematic Review 8F

1. 10, 20, 30, 40, 50, 60, 70, 80, 90, 100
2. 2, 4, 6, 8, 10, 12, 14, 16, 18, 20

3.
$$\begin{array}{r} \overset{1}{2}5 \\ +25 \\ \hline 50 \end{array}$$

4.
$$\begin{array}{r} \overset{1}{1}9 \\ +\ 4 \\ \hline 23 \end{array}$$

5.
$$\begin{array}{r} \overset{1}{3}8 \\ +15 \\ \hline 53 \end{array}$$

6.
$$\begin{array}{r} 430 \\ +223 \\ \hline 653 \end{array}$$

7.
$$\begin{array}{r} 805 \\ +192 \\ \hline 997 \end{array}$$

8.
$$\begin{array}{r} 317 \\ +651 \\ \hline 968 \end{array}$$

9. $1 + \underline{7} = 8$
10. $9 + \underline{8} = 17$
11. $4 + \underline{1} = 5$
12. $56 + 39 = 95$ pieces
13. 2, 4, $\underline{6}$ bales
14. 8 dimes = 80¢
    80¢ > 8¢
    Chance has more.
15. $2 + 2 = 4$
    $4 + 5 = 9$ hours

Lesson Practice 9A

1. 5, 10, 15, 20, 25, 30, 35, 40, 45, 50
2. 5, 10, 15, 20, 25, 30, 35, 40, 45, 50
3. 5, 10, 15, 20, 25, 30, 35, 40, 45, 50
4. 5, 10, 15, $\underline{20}$¢
5. 5, 10, 15, 20, $\underline{25}$ rooms
6. 5, 10, 15, 20, 25, 30, $\underline{35}$¢
7. 5, 10, 15, 20, 25, 30, 35, 40, $\underline{45}$ jelly beans

Lesson Practice 9B

1. 5, 10, 15, 20, 25, 30, 35, 40, 45, 50
2. 5, 10, 15, 20, 25, 30, 35, 40, 45, 50
3. 5, 10, 15, 20, 25, 30, 35, 40, 45, 50
4. 5, 10, 15, 20, $\underline{25}$¢
5. 5, 10, 15, 20, 25, $\underline{30}$ sides
6. 5, 10, 15, 20, 25, 30, 35, 40, $\underline{45}$¢
7. 5, 10, 15, 20, 25, 30, 35, $\underline{40}$ songs

Lesson Practice 9C

1. 5, 10, 15, 20, 25, 30, 35, 40, 45, 50
2. 5, 10, 15, 20, 25, 30, 35, 40, 45, 50
3. 5, 10, 15, 20, 25, 30, 35, 40, 45, 50
4. 5, 10, 15, 20, 25, 30, $\underline{35}$¢
5. 5, 10, $\underline{15}$ petals
6. 3 nickels = 15¢
   2 dimes = 20¢
   Bob has more.
7. 5, 10, 15, $\underline{20}$¢

Systematic Review 9D

1. 5, 10, 15, 20, 25, 30, 35, 40, 45, 50
2. 10, 20, 30, 40, 50, 60, 70, 80, 90, 100

3.
$$\begin{array}{r} \overset{1}{6}3 \\ +\ 7 \\ \hline 70 \end{array}$$

4.
$$\begin{array}{r} \overset{1}{2}4 \\ +48 \\ \hline 72 \end{array}$$

5.
$$\begin{array}{r} 15 \\ +44 \\ \hline 59 \end{array}$$

6.
$$\begin{array}{r} 412 \\ +216 \\ \hline 628 \end{array}$$

7.
$$\begin{array}{r} 203 \\ +302 \\ \hline 505 \end{array}$$

8.
$$\begin{array}{r} 713 \\ +272 \\ \hline 985 \end{array}$$

9. $6 + \underline{2} = 8$
10. $4 + \underline{5} = 9$
11. $7 + \underline{7} = 14$
12. 8 nickels = 40¢; yes
13. 2, 4, 6, 8, <u>10</u> feet
14. $15 + 16 = 31$ comic books
15. $6 + 2 = 8$
    $8 + 6 = 14$ passengers

Systematic Review 9E

1. 5, 10, 15, 20, 25, 30, 35, 40, 45, 50
2. 2, 4, 6, 8, 10, 12, 14, 16, 18, 20
3.
$$\begin{array}{r} \overset{1}{}27 \\ +33 \\ \hline 60 \end{array}$$
4.
$$\begin{array}{r} 81 \\ +\ 3 \\ \hline 84 \end{array}$$
5.
$$\begin{array}{r} \overset{1}{}36 \\ +14 \\ \hline 50 \end{array}$$
6.
$$\begin{array}{r} 293 \\ +104 \\ \hline 397 \end{array}$$
7.
$$\begin{array}{r} 645 \\ +321 \\ \hline 966 \end{array}$$
8.
$$\begin{array}{r} 784 \\ +215 \\ \hline 999 \end{array}$$
9. $9 + \underline{1} = 10$
10. $3 + \underline{6} = 9$
11. $7 + \underline{5} = 12$
12. 5, 10, 15, 20, 25, 30, 35, 40, 45; 9 nickels
13. $25 + 16 = 41$ times
14. $4 + 2 = 6$
    $6 + 8 = 14$ runs
15. 5 nickels = 25¢
    4 dimes = 40¢
    40¢ > 25¢; 4 dimes

Systematic Review 9F

1. 5, 10, 15, 20, 25, 30, 35, 40, 45, 50
2. 10, 20, 30, 40, 50, 60, 70, 80, 90, 100
3.
$$\begin{array}{r} 57 \\ +22 \\ \hline 79 \end{array}$$
4.
$$\begin{array}{r} \overset{1}{}74 \\ +\ 6 \\ \hline 80 \end{array}$$
5.
$$\begin{array}{r} \overset{1}{}24 \\ +18 \\ \hline 42 \end{array}$$
6.
$$\begin{array}{r} 680 \\ +119 \\ \hline 799 \end{array}$$
7.
$$\begin{array}{r} 532 \\ +222 \\ \hline 754 \end{array}$$
8.
$$\begin{array}{r} 192 \\ +207 \\ \hline 399 \end{array}$$
9. $8 + \underline{2} = 10$
10. $7 + \underline{8} = 15$
11. $9 + \underline{3} = 12$
12. dime
13. penny
14. nickel
15. $314 + 322 = 636$ dollars

Lesson Practice 10A

1. $1.66
   "one dollar and sixty-six cents"
2. $1.25
   "one dollar and twenty-five cents"
3. $2.19
   "two dollars and nineteen cents"
4. $1.30
   "one dollar and thirty cents"
5. 2 dollars, 6 dimes, and 2 pennies
   "two dollars and sixty-two cents"
6. 2 dollars and 5 pennies
   "two dollars and five cents"
7. 1 dollar, 9 dimes, and 6 pennies
   "one dollar and ninety-six cents"
8. 3 dollars, 1 dime, and 8 pennies
   "three dollars and eighteen cents"

Lesson Practice 10B

1. $3.52
   "three dollars and fifty-two cents"
2. $1.74
   "one dollar and seventy-four cents"
3. $2.46
   "two dollars and forty-six cents"
4. 4 dollars, 5 dimes, and 1 penny
   "four dollars and fifty-one cents"
5. 3 dollars and 6 dimes
   "three dollars and sixty cents"
6. 1 dollar, 8 dimes, and 1 penny
   "one dollar and eighty-one cents"
7. 2 dollars and 7 pennies
   "two dollars and seven cents"
8. two dollars and fifteen cents
   $2.15

Lesson Practice 10C

1. $4.01
   "four dollars and one cent"
2. $2.30
   "two dollars and thirty cents"

3. $1.49
   "one dollar and forty-nine cents"
4. 3 dollars, 2 dimes, and 3 pennies
   "three dollars and twenty-three cents"
5. 1 dollar and 8 pennies
   "one dollar and eight cents"
6. 4 dollars, 5 dimes, and 2 pennies
   "four dollars and fifty-two cents"
7. 2 dollars and 9 dimes
   "two dollars and ninety cents"
8. six dollars and seventy-three cents
   $6.73

Systematic Review 10D

1. $2.67
   "two dollars and sixty-seven cents"
2. 1 dollar, 4 dimes, and 8 pennies
   "one dollar and forty-eight cents"
3. 2 dollars, 7 dimes, and 3 pennies
   "two dollars and seventy-three cents"
4. 4 dollars and 5 pennies
   "four dollars and five cents"
5. 3 dollars and 6 dimes
   "three dollars and sixty cents"
6. 5, 10, 15, 20, 25¢
7.
$$\begin{array}{r} \overset{1}{}4\,9 \\ +\ \ 9 \\ \hline 5\,8 \end{array}$$
8.
$$\begin{array}{r} 3\,1\,1 \\ +2\,3\,8 \\ \hline 5\,4\,9 \end{array}$$
9.
$$\begin{array}{r} \overset{1}{}6\,5 \\ +2\,5 \\ \hline 9\,0 \end{array}$$
10. $7 + \underline{6} = 13$
11. $3 + \underline{2} = 5$
12. $4 + \underline{8} = 12$
13. five dollars and four cents = $5.04
14. $19 + 25 = 44$ rides
15. $4 + 4 = 8$
    $8 + 9 = 17$ cards

Systematic Review 10E

1. $3.20
   "three dollars and twenty cents"
2. 2 dollars, 3 dimes, and 1 penny
   "two dollars and thirty-one cents"
3. 4 dollars, 5 dimes, and 5 pennies
   "four dollars and fifty-five cents"
4. 1 dollar and 6 pennies
   "one dollar and six cents"
5. 3 dollars, 7 dimes, and 8 pennies
   "three dollars and seventy-eight cents"
6. 10, 20, 30, 40, 50, 60, 70, 80¢
7. 17
   +18
   35
8. 555
   + 132
   687
9. 49
   +34
   83
10. 80
11. 10
12. 30
13. 5, 10, 15, 20, 25, 30,
    35, 40, 45¢ = $0.45
14. 5 + 2 = 7
    7 + 2 = 9 roses
15. 30¢

Systematic Review 10F

1. $1.08
   "one dollar and eight cents"
2. 1 dollar, 1 dime, and 6 pennies
   "one dollar and sixteen cents"
3. 3 dollars and 9 pennies
   "three dollars and nine cents"
4. 2 dollars, 6 dimes, and 5 pennies
   "two dollars and sixty-five cents"
5. 4 dollars and 7 dimes
   "four dollars and seventy cents"

6. 5, 10, 15¢
7. 92
   + 4
   96
8. 337
   +202
   539
9. 61
   +29
   90
10. 10 > 7
11. 8 = 8
12. 27 < 72
13. $8.69
14. 3 + 4 = 7
    7 + 8 = 15 miles
15. 29 + 18 = 47 miles

Lesson Practice 11A

1. 200
2. 200
3. 400
4. done
5. (600) 628
   +(200) +175
   (800   803
6. (400) 359
   +(300) +254
   (700) 613
7. (500) 537
   +(200) +233
   (700) 770
8. (200) 168
   +(500) +452
   (700) 620
9. (100) 123
   +(100) + 88
   (200) 211

10.
```
            1 1
   (700)    676
  +(100)   +145
  (800)     821
```

11.
```
            1 1
   (300)    299
  +(300)   +311
  (600)     610
```

12.   124 + 176 = 300 lights

Lesson Practice 11B

1.   500
2.   500
3.   600

4.
```
             1
   (400)    359
  +(100)   +126
  (500)     485
```

5.
```
             1
   (100)    138
  +(200)   +212
  (300)     350
```

6.
```
   (200)    157
  +(100)   +142
  (300)     299
```

7.
```
             1
   (200)    227
  +(000)   + 39
  (200)     266
```

8.
```
             1
   (400)    449
  +(100)   +137
  (500)     586
```

9.
```
             1
   (200)    235
  +(100)   +145
  (300)     380
```

10.
```
             1
   (100)    109
  +(200)   +207
  (300)     316
```

11.
```
            1
   (400)    416
  +(300)   +329
  (700)     745
```

12.   123 + 169 = 292 pages

Lesson Practice 11C

1.   500
2.   100
3.   300

4.
```
             1
   (200)    217
  +(300)   +324
  (500)     541
```

5.
```
             1
   (300)    266
  +(000)   + 18
  (300)     284
```

6.
```
   (100)    134
  +(400)   +365
  (500)     499
```

7.
```
             1
   (100)    119
  +(200)   +207
  (300)     326
```

8.
```
            1 1
   (600)    555
  +(300)   +348
  (900)     903
```

9.
```
             1
   (800)    806
  +(100)   +106
  (900)     912
```

10.
```
             1
   (100)    119
  +(200)   +217
  (300)     336
```

11.
```
            1 1
   (200)    248
  +(300)   +252
  (500)     500
```

12.   263 + 179 = 442 miles

Systematic Review 11D

1. 800
2. 100

3.
$$\begin{array}{r} \overset{1}{} \\ 806 \\ +106 \\ \hline 912 \end{array}$$

4.
$$\begin{array}{r} \overset{11}{} \\ 248 \\ +252 \\ \hline 500 \end{array}$$

5.
$$\begin{array}{r} \overset{1}{} \\ 337 \\ +172 \\ \hline 509 \end{array}$$

6.
$$\begin{array}{r} \overset{1}{} \\ 54 \\ +28 \\ \hline 82 \end{array}$$

7.
$$\begin{array}{r} \overset{1}{} \\ 53 \\ +37 \\ \hline 90 \end{array}$$

8.
$$\begin{array}{r} \overset{1}{} \\ 18 \\ +29 \\ \hline 47 \end{array}$$

9. $1 - 1 = 0$
10. $10 - 2 = 8$
11. $8 - 1 = 7$
12. $3 - 0 = 3$
13. $4 - 3 = 1$
14. $6 - 2 = 4$
15. $5 - 4 = 1$
16. $8 - 2 = 6$
17. 2, 4, 6, 8, 10, 12, 14, 16, 18, 20
18. $5.26
19. $55 + 78 = 133$ miles
20. $145 + $56 = 201

Systematic Review 11E

1. 400
2. 200

3.
$$\begin{array}{r} \overset{11}{} \\ 235 \\ +365 \\ \hline 600 \end{array}$$

4.
$$\begin{array}{r} 300 \\ +409 \\ \hline 709 \end{array}$$

5.
$$\begin{array}{r} \overset{1}{} \\ 249 \\ +132 \\ \hline 381 \end{array}$$

6.
$$\begin{array}{r} \overset{1}{} \\ 28 \\ +38 \\ \hline 66 \end{array}$$

7.
$$\begin{array}{r} \overset{1}{} \\ 65 \\ +35 \\ \hline 100 \end{array}$$

8.
$$\begin{array}{r} \overset{1}{} \\ 58 \\ +42 \\ \hline 100 \end{array}$$

9. $4 - 2 = 2$
10. $7 - 2 = 5$
11. $3 - 1 = 2$
12. $11 - 2 = 9$
13. $6 - 5 = 1$
14. $8 - 0 = 8$
15. $10 - 9 = 1$
16. $9 - 2 = 7$
17. 5, 10, 15, 20, 25, 30, 35, 40, 45, 50
18. $17 + 9 = 26$ inches
19. $138 + 256 = 394$ penguins
20. $5 - 2 = 3$ eggs

Systematic Review 11F

1. 500
2. 700

3.
$$\begin{array}{r} \overset{1}{} \\ 429 \\ +266 \\ \hline 695 \end{array}$$

4.
$$\begin{array}{r} {}^{1} \\ 10\,1 \\ +\ 89 \\ \hline 190 \end{array}$$

5.
$$\begin{array}{r} {}^{1} \\ 238 \\ +243 \\ \hline 481 \end{array}$$

6.
$$\begin{array}{r} {}^{1} \\ 92 \\ +\ 8 \\ \hline 100 \end{array}$$

7.
$$\begin{array}{r} {}^{1} \\ 48 \\ +32 \\ \hline 80 \end{array}$$

8.
$$\begin{array}{r} {}^{1} \\ 63 \\ +27 \\ \hline 90 \end{array}$$

9. $5-3=2$
10. $10-2=8$
11. $7-5=2$
12. $6-2=4$
13. $9-7=2$
14. $8-2=6$
15. $10-8=2$
16. $11-9=2$
17. 10, 20, 30, 40, 50, 60, 70, 80, 90, 100
18. $(20)+(50)=(70)$ pieces
19. $(500)+(300)=(800)$
    $512+345=857$ miles
20. $\$0.08-\$0.06=\$0.02$

Lesson Practice 12A
1. done

2.
$$\begin{array}{r} {}^{1\ 1} \\ \$7.09 \\ +1.92 \\ \hline \$9.01 \end{array}$$

3.
$$\begin{array}{r} \$3.33 \\ +1.44 \\ \hline \$4.77 \end{array}$$

4.
$$\begin{array}{r} {}^{1} \\ \$6.50 \\ +2.77 \\ \hline \$9.27 \end{array}$$

5.
$$\begin{array}{r} \$4.00 \\ +2.51 \\ \hline \$6.51 \end{array}$$

6.
$$\begin{array}{r} {}^{1} \\ \$5.19 \\ +1.38 \\ \hline \$6.57 \end{array}$$

7.
$$\begin{array}{r} \$1.00 \\ +0.75 \\ \hline \$1.75 \end{array}$$

8.
$$\begin{array}{r} \$2.03 \\ +1.90 \\ \hline \$3.93 \end{array}$$

9.
$$\begin{array}{r} {}^{1} \\ \$8.75 \\ +0.80 \\ \hline \$9.55 \end{array}$$

10. $\$5.25+\$3.38=\$8.63$
11. $\$2.63+\$5.50=\$8.13$
12. $\$2.99+\$3.61=\$6.60$

Lesson Practice 12B
1.
$$\begin{array}{r} {}^{1} \\ \$7.65 \\ +0.60 \\ \hline \$8.25 \end{array}$$

2.
$$\begin{array}{r} {}^{1} \\ \$6.31 \\ +1.29 \\ \hline \$7.60 \end{array}$$

3.
$$\begin{array}{r} {}^{1} \\ \$5.83 \\ +0.24 \\ \hline \$6.07 \end{array}$$

4.
$$\begin{array}{r} {}^{1} \\ \$3.19 \\ +0.90 \\ \hline \$4.09 \end{array}$$

5.
$$\begin{array}{r} \$2.00 \\ +0.98 \\ \hline \$2.98 \end{array}$$

6.
```
  $1.03
 +1.25
 ─────
  $2.28
```

7.
```
     1
  $3.72
 +4.08
 ─────
  $7.80
```

8.
```
   1 1
  $1.99
 +1.82
 ─────
  $3.81
```

9.
```
   1 1
  $2.87
 +6.89
 ─────
  $9.76
```

10. $3.45 + $1.99 = $5.44

11. $5.55 + $2.15 = $7.70

12. $6.34 + $2.95 = $9.29

Lesson Practice 12C

1.
```
     1
  $2.13
 +1.92
 ─────
  $4.05
```

2.
```
     1
  $4.71
 +1.36
 ─────
  $6.07
```

3.
```
     1
  $6.41
 +0.39
 ─────
  $6.80
```

4.
```
  $5.00
 +2.50
 ─────
  $7.50
```

5.
```
     1
  $6.63
 +2.44
 ─────
  $9.07
```

6.
```
     1
  $7.35
 +1.05
 ─────
  $8.40
```

7.
```
     1
  $1.63
 +0.72
 ─────
  $2.35
```

8.
```
   1 1
  $4.99
 +3.79
 ─────
  $8.78
```

9.
```
     1
  $6.33
 +2.91
 ─────
  $9.24
```

10. $5.10 + $3.91 = $9.01

11. $2.50 + $4.50 = $7.00

12. $3.72 + $3.68 = $7.40

$7.40 > $6.00; yes

Systematic Review 12D

1.
```
     1
  $1.66
 +4.08
 ─────
  $5.74
```

2.
```
     1
  $3.09
 +2.56
 ─────
  $5.65
```

3.
```
     1
  $3.57
 +2.62
 ─────
  $6.19
```

4.
```
  1 1
   422
 +389
 ────
  811
```

5.
```
    1
   19
 +16
 ───
   35
```

6.
```
    1
   17
 +25
 ───
   42
```

7. 12 − 9 = 3

8. 18 − 9 = 9

9. 9 − 9 = 0

10. 14 − 9 = 5

11. $17 - 9 = 8$
12. $13 - 9 = 4$
13. $16 - 9 = 7$
14. $15 - 9 = 6$
15. $2.35
   "two dollars and thirty-five cents"
16. $G - 9 = 2$; $G = 11$ gifts
17. $(50) + (60) = (110)$
   $46 + 63 = 109$ chairs
18. $2.78 + $1.19 = $3.97

15. $1.26
   "one dollar and twenty-six cents"
16. $D - 8 = 5$
   $D = 13$ dimes
17. $(300) + (200) = (500)$
   $267 + 197 = 464$ candies
18. 3 dimes = 30¢
   5 nickels = 25¢
   30¢ + 25¢ = 55¢
   $0.55

Systematic Review 12E

1.
```
  1 1
$1.68
+4.77
─────
$6.45
```

2.
```
  1 1
$4.56
+4.44
─────
$9.00
```

3.
```
   1
$2.63
+0.51
─────
$3.14
```

4.
```
   1
 684
+122
────
 806
```

5.
```
  1
 62
+29
───
 91
```

6.
```
  1
 83
+ 7
───
 90
```

7. $11 - 8 = 3$
8. $9 - 8 = 1$
9. $17 - 8 = 9$
10. $12 - 8 = 4$
11. $14 - 8 = 6$
12. $13 - 8 = 5$
13. $15 - 8 = 7$
14. $16 - 8 = 8$

Systematic Review 12F

1.
```
  1 1
$2.78
+6.58
─────
$9.36
```

2.
```
   1
$3.52
+1.77
─────
$5.29
```

3.
```
$8.91
+0.05
─────
$8.96
```

4.
```
  1 1
 379
+264
────
 643
```

5.
```
  1
 54
+18
───
 72
```

6.
```
  1
 47
+ 9
───
 56
```

7. $12 - 6 = 6$
8. $13 - 8 = 5$
9. $10 - 5 = 5$
10. $14 - 7 = 7$
11. $12 - 8 = 4$
12. $17 - 8 = 9$
13. $8 - 4 = 4$
14. $6 - 3 = 3$
15. $0.54
   "fifty-four cents"

16. B − 8 = 8; B = 16 birds
17. 3 nickels = 15¢
    15¢ + 5¢ = 20¢
    $0.20
18. 35 + 17 = 52 minutes

Lesson Practice 13A
1. 2 + 8 + 5 = 15
2. 6 + 3 + 4 = 13
3. 5 + 4 + 5 = 14
4. 8 + 2 + 6 + 1 = 17
5. 6 + 2 + 4 + 7 = 19
6. 3 + 7 + 5 + 5 = 20
7. 40 + 50 + 10 = 100
8. 30 + 70 + 40 = 140
9. 51 + 59 + 20 = 130
10. 9 + 1 + 8 + 2 = 20
11. 6 + 1 + 8 + 4 + 2 = 21
12. 4 + 5 + 6 + 5 = 20 presents

Lesson Practice 13B
1. 6 + 4 + 9 = 19
2. 9 + 5 + 5 = 19
3. 7 + 2 + 3 = 12
4. 8 + 5 + 4 + 2 = 19
5. 9 + 1 + 7 + 3 = 20
6. 1 + 2 + 3 + 9 = 15
7. 60 + 40 + 20 = 120
8. 20 + 60 + 80 = 160
9. 43 + 27 + 61 = 131
10. 8 + 2 + 3 + 7 + 2 = 22
11. 4 + 9 + 6 + 2 + 1 = 22
12. 8 + 2 + 3 + 3 = 16 laps

Lesson Practice 13C
1. 2 + 8 + 7 = 17
2. 5 + 1 + 5 = 11
3. 4 + 3 + 3 = 10
4. 2 + 8 + 2 + 3 = 15
5. 3 + 8 + 7 + 2 = 20
6. 9 + 5 + 6 + 1 = 21

7. 80 + 50 + 10 + 10 = 150
8. 40 + 60 + 4 + 2 = 106
9. 84 + 26 + 17 + 23 = 150
10. 6 + 7 + 3 + 4 + 6 = 26
11. 5 + 5 + 4 + 3 + 7 = 24
12. 11 + 14 + 5 + 10 + 4 = 44 books

Systematic Review 13D
1. 3 + 7 + 6 = 16
2. 3 + 9 + 1 + 7 = 20
3. 42 + 64 + 82 + 2 = 190
4.
$$\begin{array}{r} {\scriptstyle 1\ 1} \\ \$2.85 \\ +6.56 \\ \hline \$9.41 \end{array}$$
5.
$$\begin{array}{r} {\scriptstyle 1\ 1} \\ 149 \\ +273 \\ \hline 422 \end{array}$$
6.
$$\begin{array}{r} {\scriptstyle 1} \\ 14 \\ +\ 9 \\ \hline 23 \end{array}$$
7. 10 − 8 = 2
8. 10 − 4 = 6
9. 9 − 6 = 3
10. 10 − 7 = 3
11. 9 − 2 = 7
12. 9 − 4 = 5
13. 9 − 1 = 8
14. 10 − 6 = 4
15. 10, 20, 30, 40, 50, 60, 70, 80, 90, 100
16. triangle
17. 13 + 15 + 11 = 39 patients
18. $6.18 + $2.00 = $8.18
    $8.18 < $9.00; no

Systematic Review 13E
1. 2 + 4 + 2 = 8
2. 6 + 3 + 8 + 2 = 19
3. 10 + 10 + 20 + 20 = 60

4. $\begin{array}{r} {\scriptstyle 1\ 1} \\ \$3.46 \\ +2.54 \\ \hline \$6.00 \end{array}$

5. $\begin{array}{r} 100 \\ +278 \\ \hline 378 \end{array}$

6. $\begin{array}{r} {\scriptstyle 1} \\ 79 \\ +88 \\ \hline 167 \end{array}$

7. $10 - 2 = 8$

8. $10 - 5 = 5$

9. $9 - 5 = 4$

10. $9 - 7 = 2$

11. $10 - 3 = 7$

12. $9 - 8 = 1$

13. $9 - 3 = 6$

14. $10 - 4 = 6$

15.   2, 4, 6, 8, 10, 12, 14, 16, 18, 20

16.   square, rectangle

17. $35 + 22 + 12 = 69$ cones

18.   9 nickels = 45¢
      45¢ + 7¢ = 52¢
      $0.52

Systematic Review 13F

1. $5 + 5 + 2 = 12$

2. $5 + 6 + 4 + 7 = 22$

3. $23 + 37 + 23 + 12 = 95$

4. $\begin{array}{r} \$4.36 \\ +1.22 \\ \hline \$5.58 \end{array}$

5. $\begin{array}{r} {\scriptstyle 1} \\ 116 \\ +\ 68 \\ \hline 184 \end{array}$

6. $\begin{array}{r} {\scriptstyle 1} \\ 16 \\ +44 \\ \hline 60 \end{array}$

7. $17 - 8 = 9$

8. $14 - 7 = 7$

9. $10 - 3 = 7$

10. $9 - 5 = 4$

11. $15 - 9 = 6$

12. $9 - 7 = 2$

13. $12 - 8 = 4$

14. $16 - 9 = 7$

15.   5, 10, 15, 20, 25, 30, 35, 40, 45, 50

16.
17.
18.

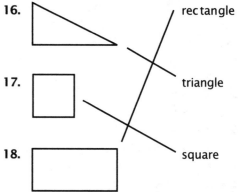

A square is also a rectangle, so the student may correctly draw another line connecting the square to the word "rectangle."

19.   6 nickels = 30¢
      5 dimes = 50¢
      30¢ + 50¢ + 4¢ = 84¢
      $0.84

20. $33 + 49 = 82$ raisins

Lesson Practice 14A

1.   sides: 2"
      base: 3"

2.   4"

3.   5"

4.   3"; line 3; 5 - 3 = 2

5.   sides: 2"
      top and bottom: 3"

6.   1 foot

Lesson Practice 14B

1.   all sides: 1"

2.   2"

3.   7"

4.   6"; line 3; 7 - 6 = 1

5.   all sides: 2"

6.   12

Lesson Practice 14C

1. sides: 1"
   top and bottom: 3"
2. 5"
3. 1"
4. 4"; line 4; 4 - 1 = 3
5. all sides: 3"
6. 12"+12" = 24"

Systematic Review 14D

1. 3"
2. 7"; line 2; 7 - 3 = 4

3.
$$\begin{array}{r} \overset{1\ 1}{} \\ \$1.77 \\ +2.78 \\ \hline \$4.55 \end{array}$$

4.
$$\begin{array}{r} \overset{1}{} \\ 118 \\ +122 \\ \hline 240 \end{array}$$

5.
$$\begin{array}{r} 23 \\ +24 \\ \hline 47 \end{array}$$

6. $1+2+3+7+8+9 = 30$
7. $6+6+4+3+7 = 26$
8. $6 = 6$
9. $6 < 8$
10. $7 - 3 = 4$
11. $8 - 5 = 3$
12. $7 - 4 = 3$
13. $8 - 3 = 5$
14. $9 - 5 = 4$
15. $15 - 9 = 6$
16. three
17. $6 - 3 = 3$ dimes
    3 dimes = 30¢ or $0.30
18. $6+5+10+4 = 25$ pieces

Systematic Review 14E

1. 6"
2. 2"; line 1; 6 - 2 = 4

3.
$$\begin{array}{r} \overset{1\ 1}{} \\ \$3.63 \\ +2.77 \\ \hline \$6.40 \end{array}$$

4.
$$\begin{array}{r} \overset{1}{} \\ 452 \\ +181 \\ \hline 633 \end{array}$$

5.
$$\begin{array}{r} \overset{1}{} \\ 89 \\ +87 \\ \hline 176 \end{array}$$

6. $3+3+7+3+2 = 18$
7. $10+2+10+2 = 24$
8. $4 > 3$
9. $11 - 7 = 4$
10. $13 - 7 = 6$
11. $8 - 3 = 5$
12. $16 - 7 = 9$
13. $12 - 7 = 5$
14. $15 - 7 = 8$
15. four
16. $256 + 289 = 545$ miles
17. $3+4+1+2 = 10$ calls
18. $12+12+12 = 36"$

Systematic Review 14F

1. 4"
2. 5", line 2, 5 - 4 = 1

3.
$$\begin{array}{r} \overset{1}{} \\ \$1.56 \\ +2.38 \\ \hline \$3.94 \end{array}$$

4.
$$\begin{array}{r} \overset{1}{} \\ 419 \\ +419 \\ \hline 838 \end{array}$$

5.
$$\begin{array}{r} \overset{1}{} \\ 39 \\ +42 \\ \hline 81 \end{array}$$

6. $5+5+5+5 = 20$
7. $3+5+10 = 18$
8. $5 < 8$
9. $11 - 6 = 5$

10. $14 - 6 = 8$
11. $15 - 6 = 9$
12. $13 - 6 = 7$
13. $15 - 7 = 8$
14. $7 - 4 = 3$
15. four
16. 7 dimes = 70¢
    3 nickels = 15¢
    70¢ + 15¢ = 85¢ or $0.85
17. $3 + 3 = 6$ pages read
    $13 - 6 = 7$ pages left
18. $56 + 35 = 91$ miles

Lesson Practice 15A

1. done
2. square (a type of rectangle)
   $6 + 6 + 6 + 6 = 24"$
3. triangle
   $3 + 4 + 5 = 12"$
4. rectangle
   $7 + 2 + 7 + 2 = 18"$

Lesson Practice 15B

1. rectangle
   $3 + 8 + 3 + 8 = 22"$
2. square (a type of rectangle)
   $2 + 2 + 2 + 2 = 8"$
3. triangle
   $6 + 8 + 10 = 24"$
4. rectangle
   $4 + 6 + 4 + 6 = 20"$

Lesson Practice 15C

1. square (a type of rectangle)
   $9 + 9 + 9 + 9 = 36"$
2. rectangle
   $2 + 5 + 2 + 5 = 14"$
3. square (a type of rectangle)
   $7 + 7 + 7 + 7 = 28"$
4. triangle
   $4 + 7 + 10 = 21"$

Systematic Review 15D

1. triangle
   $6 + 6 + 7 = 19"$
2. rectangle
   $10 + 15 + 10 + 15 = 50"$
3. $\begin{array}{r} 1 \\ \$2.49 \\ +1.32 \\ \hline \$3.81 \end{array}$
4. $\begin{array}{r} 300 \\ +409 \\ \hline 709 \end{array}$
5. $\begin{array}{r} 1 \\ 27 \\ +25 \\ \hline 52 \end{array}$
6. $11 - 5 = 6$
7. $12 - 4 = 8$
8. $13 - 5 = 8$
9. $11 - 3 = 8$
10. $12 - 5 = 7$
11. $11 - 4 = 7$
12. $10 - 7 = 3$ years
13. $53 + 8 = 61$ papers
14. $12 + 12 = 24"$
15. 9 nickels = 45¢
    1 dime = 10¢
    45¢ + 10¢ = 55¢ or $0.55

Systematic Review 15E

1. square (a type of rectangle)
   $6 + 6 + 6 + 6 = 24"$
2. rectangle
   $5 + 12 + 5 + 12 = 34"$
3. $\begin{array}{r} 1 \\ \$4.28 \\ +1.65 \\ \hline \$5.93 \end{array}$
4. $\begin{array}{r} 1\,1 \\ 285 \\ +156 \\ \hline 441 \end{array}$

5.
```
   1
   45
 +55
  100
```

6.
7	9	16
3	5	8
10	14	24

7.
6	4	10
9	8	17
15	12	27

8. $14 - 5 = 9$
9. $12 - 3 = 9$
10. $6 - 1 = 5$
11. $13 - 4 = 9$
12. $4 + 4 + 4 + 4 = 16"$
13. $31 + 31 + 28 = 90$ days
14. $12 + 12 + 12 + 12 = 48"$

Systematic Review 15F
1. triangle
   $7 + 8 + 9 = 24"$
2. rectangle
   $11 + 18 + 11 + 18 = 58"$
3.
```
   1
 $2.48
 +1.29
 $3.77
```
4.
```
  11
  183
 + 77
  260
```
5.
```
   63
  +26
   89
```
6. 2, 4, 6, 8, 10, 12, 14, 16, 18, 20
7. 5, 10, 15, 20, 25, 30, 35, 40, 45, 50
8. $9 - 6 = 3$
9. $11 - 7 = 4$
10. $12 - 5 = 7$
11. $13 - 9 = 4$
12. $11 - 5 = 6$ guests
13. $6 + 8 + 10 = 24'$ (24 feet)
14. $9 + 9 = 18$ runs

Lesson Practice 16A
1. done
2. 35,361
   "thirty-five thousand,
   three hundred sixty-one"
3. 785,892
   "seven hundred eighty-five thousand,
   eight hundred ninety-two"
4. 265,143
   "two hundred sixty-five thousand,
   one hundred forty-three"
5. 6,237
   "six thousand, two hundred thirty-seven"
6. done
7. $2,000 + 300 + 50 + 6$
8. done
9. 1,542
10. done
11.
```
    1
    915
  +436
  1,351
```
12.
```
    1
    381
  +727
  1,108
```

Lesson Practice 16B
1. 2,794
   "two thousand, seven hundred ninety-four"
2. 16,322
   "sixteen thousand,
   three hundred twenty-two"
3. 651,741
   "six hundred fifty-one thousand,
   seven hundred forty-one"
4. 536,583
   "five hundred thirty-six thousand,
   five hundred eighty-three"
5. 2,549
   "two thousand, five hundred forty-nine"
6. $40,000 + 1,000 + 400 + 50 + 6$
7. $200,000 + 30,000 + 8,000 + 100 + 90 + 9$
8. 3,121

9. 45,616

10.
$$
\begin{array}{r}
\overset{1}{5\,9\,3} \\
+5\,5\,1 \\
\hline
1,144
\end{array}
$$

11.
$$
\begin{array}{r}
\overset{1}{8\,7\,6} \\
+4\,3\,1 \\
\hline
1,307
\end{array}
$$

12.
$$
\begin{array}{r}
9\,6\,7 \\
+2\,0\,2 \\
\hline
1,169
\end{array}
$$

Lesson Practice 16C

1. 1,224
   "one thousand,
   two hundred twenty-four"
2. 43,638
   "forty-three thousand,
   six hundred thirty-eight"
3. 247,500
   "two hundred forty-seven thousand,
   five hundred"
4. 122,472
   "one hundred twenty-two thousand,
   four hundred seventy-two"
5. 7,294
   "seven thousand,
   two hundred ninety-four"
6. $50,000 + 6,000 + 600 + 40 + 4$
7. $3,000 + 200 + 50 + 6$
8. 1,838
9. 33,230
10.
$$
\begin{array}{r}
4\,5\,3 \\
+7\,1\,4 \\
\hline
1,167
\end{array}
$$

11.
$$
\begin{array}{r}
\overset{1\ 1}{3\,4\,5} \\
+9\,7\,8 \\
\hline
1,323
\end{array}
$$

12.
$$
\begin{array}{r}
7\,1\,6 \\
+5\,6\,3 \\
\hline
1,279
\end{array}
$$

Systematic Review 16D

1. 4,819
   "four thousand,
   eight hundred nineteen"
2. 57,284
   "fifty-seven thousand,
   two hundred eighty-four"
3. rectangle
   $11 + 32 + 11 + 32 = 86"$
4.
$$
\begin{array}{r}
9\,0\,1 \\
+8\,5\,0 \\
\hline
1,751
\end{array}
$$

5.
$$
\begin{array}{r}
\overset{1}{5\,3\,4} \\
+6\,7\,3 \\
\hline
1,207
\end{array}
$$

6.
$$
\begin{array}{r}
\$4.\,12 \\
+4.\,71 \\
\hline
\$8.83
\end{array}
$$

7. $17 + 23 + 55 = 95$
8. $10 - 6 = 4$
9. $15 - 8 = 7$
10. $14 - 6 = 8$
11. $6 - 3 = 3$
12. $12 - 7 = 5$
13. $6 - 4 = 2$
14. $\$48 + \$32 + \$21 = \$101$
15. $12 + 12 + 12 = 36"$
16. $16 - 8 = 8$ years

Systematic Review 16E

1. 6,211
   "six thousand, two hundred eleven"
2. 28,616
   "twenty-eight thousand,
   six hundred sixteen"
3. triangle
   $11 + 12 + 13 = 36"$
4.
$$
\begin{array}{r}
\overset{1}{5\,5\,9} \\
+5\,2\,4 \\
\hline
1,083
\end{array}
$$

5.
$$\begin{array}{r} 1 \\ 943 \\ +475 \\ \hline 1,418 \end{array}$$

6.
$$\begin{array}{r} 1 \\ \$\ 3.37 \\ +\ 8.29 \\ \hline \$11.66 \end{array}$$

7. $65+28+45=138$

8. $9-3=6$

9. $8-5=3$

10. $14-8=6$

11. $11-4=7$

12. $16-7=9$

13. $7-1=6$

14. $\$(200)+\$(200)=\$(400)$
$\$198+\$242=\$440$

15. 8 dimes = 80¢
80¢ – 30¢ = 50¢ or $0.50

16. $9+7+9+7=32'$

Systematic Review 16F

1. 7,854
"seven thousand, eight hundred fifty-four"

2. 815,231

3. square (a type of rectangle)
$6+6+6+6=24"$

4.
$$\begin{array}{r} 624 \\ +413 \\ \hline 1,037 \end{array}$$

5.
$$\begin{array}{r} 426 \\ +873 \\ \hline 1,299 \end{array}$$

6.
$$\begin{array}{r} 1 \\ \$1.32 \\ +3.38 \\ \hline \$4.70 \end{array}$$

7. $71+53+11=135$

8. $8-7=1$

9. $11-8=3$

10. $14-9=5$

11. $9-4=5$

12. $13-5=8$

13. $8-3=5$

14. $(500)+(500)=(1,000)$
$512+471=983$ cars

15. $10-7=3$ apples

16. 5 nickels = 25¢
4 dimes = 40¢
40¢ > 25¢; 4 dimes

Lesson Practice 17A

1. 5,000

2. 1,000

3. 90,000

4. 50,000

5. done

6.
$$\begin{array}{rr} 1\ 1\ 1 & \\ 4,859 & (5,000) \\ +2,444 & +(2,000) \\ \hline 7,303 & (7,000) \end{array}$$

7.
$$\begin{array}{rr} 1 & \\ 9,253 & (9,000) \\ +7,845 & +\ (8,000) \\ \hline 17,098 & (17,000) \end{array}$$

8.
$$\begin{array}{rr} 1 & \\ 7,132 & (7,000) \\ +1,186 & +(1,000) \\ \hline 8,318 & (8,000) \end{array}$$

9.
$$\begin{array}{rr} 1\quad\ 1 & \\ 3,624 & (4,000) \\ +4,418 & +(4,000) \\ \hline 8,042 & (8,000) \end{array}$$

10.
$$\begin{array}{rr} 1\ 1\ 1 & \\ 2,852 & (3,000) \\ +3,149 & +(3,000) \\ \hline 6,001 & (6,000) \end{array}$$

11. $3,152+7,321=10,473$ miles

12. $5,232+3,765=8,997$ fish

Lesson Practice 17B

1. 7,000

2. 4,000

3. 50,000

4. 20,000

5.
$$\begin{array}{r}
1\ 1\\
5,242\\
+3,765\\
\hline
9,007
\end{array}$$
(5,000)
+(4,000)
(9,000)

6.
$$\begin{array}{r}
1\\
9,287\\
+1,321\\
\hline
10,608
\end{array}$$
(9,000)
+ (1,000)
(10,000)

7.
$$\begin{array}{r}
1\ 1\\
6,463\\
+8,765\\
\hline
15,228
\end{array}$$
(6,000)
+ (9,000)
(15,000)

8.
$$\begin{array}{r}
1\\
7,214\\
+1,108\\
\hline
8,322
\end{array}$$
(7,000)
+(1,000)
(8,000)

9.
$$\begin{array}{r}
1\ 1\\
2,817\\
+9,236\\
\hline
12,053
\end{array}$$
(3,000)
+ (9,000)
(12,000)

10.
$$\begin{array}{r}
1\ 1\\
3,680\\
+3,384\\
\hline
7,064
\end{array}$$
(4,000)
+(3,000)
(7,000)

11. 5,740 + 4,291 = 10,031 plants
12. 4,987 + 3,732 = 8,719 seeds

Lesson Practice 17C

1. 1,000
2. 6,000
3. 80,000
4. 10,000

5.
$$\begin{array}{r}
9,413\\
+1,245\\
\hline
10,658
\end{array}$$
(9,000)
+ (1,000)
(10,000)

6.
$$\begin{array}{r}
1\\
9,287\\
+7,491\\
\hline
16,778
\end{array}$$
(9,000)
+ (7,000)
(16,000)

7.
$$\begin{array}{r}
1\ 1\ 1\\
5,486\\
+4,528\\
\hline
10,014
\end{array}$$
(5,000)
+ (5,000)
(10,000)

8.
$$\begin{array}{r}
9,025\\
+3,354\\
\hline
12,379
\end{array}$$
(9,000)
+ (3,000)
(12,000)

9.
$$\begin{array}{r}
7,513\\
+7,254\\
\hline
14,767
\end{array}$$
(8,000)
+ (7,000)
(15,000)

10.
$$\begin{array}{r}
1\ 1\\
1,890\\
+3,672\\
\hline
5,562
\end{array}$$
(2,000)
+(4,000)
(6,000)

11. $5,486 + $1,194 = $6,680
12. 8,972 + 9,221 = 18,193 people

Systematic Review 17D

1. 3,000
2. 1,000

3.
$$\begin{array}{r}
1\ 1\ 1\\
6,788\\
+2,467\\
\hline
9,255
\end{array}$$

4.
$$\begin{array}{r}
1\ 1\\
2,355\\
+1,672\\
\hline
4,027
\end{array}$$

5.
$$\begin{array}{r}
\$\ 8.42\\
+\ 3.21\\
\hline
\$11.63
\end{array}$$

6. 3,188
"three thousand, one hundred eighty-eight"
7. 8 + 16 + 8 + 16 = 48"
8. 14 − 7 = 7
9. 11 − 6 = 5
10. 7 − 3 = 4
11. 12 − 3 = 9
12. 9 − 5 = 4
13. 8 − 4 = 4
14. 5 + 5 = 10 eggs picked up
10 − 3 = 7 eggs left
15. 296 + 316 = 612 feet
16. $16 + $18 + $9 = $43

Systematic Review 17E

1. 40,000
2. 80,000
3.
```
   1
  1,476
 +7,813
  9,289
```
4.
```
  1,621
 +4,157
  5,778
```
5.
```
   1
 $7.16
 +2.79
 $9.95
```
6. 5,400
   "five thousand, four hundred"
7. $15 + 15 + 15 + 15 = 60'$
8. $12 - 4 = 8$
9. $10 - 3 = 7$
10. $7 - 4 = 3$
11. $3 - 2 = 1$
12. $15 - 6 = 9$
13. $9 - 3 = 6$
14. 3 dimes = 30¢
    30¢ + 4¢ = 34¢ or $0.34
15. $75 + 108 = 183$ items
16. $1,575 + 1,892 = 3,467$ mosquitoes

Systematic Review 17F

1. 1,000
2. 9,000
3.
```
    1
  7,438
 +9,114
 16,552
```
4.
```
    1
  6,408
 +4,379
 10,787
```
5.
```
    1
 $2.56
 +1.24
 $3.80
```

6. 131,528
   "one hundred thirty-one thousand,
   five hundred twenty-eight"
7. $9 + 12 + 15 = 36'$
8. $13 - 4 = 9$
9. $11 - 3 = 8$
10. $14 - 5 = 9$
11. $10 - 9 = 1$
12. $6 - 0 = 6$
13. $8 - 8 = 0$
14. $12 - 7 = 5$ people
15. $(200) + (300) = (500)$
    $235 + 256 = 491$ miles
16. $10 + 12 + 18 = 40$ flowers

Lesson Practice 18A

1. done
2.
```
   22
   294    (300)
   187    (200)
   306    (300)
  +813  + (800)
  1,600   (1,600)
```
3.
```
   22
   493    (500)
   215    (200)
   485    (500)
   324    (300)
  +106  + (100)
  1,623   (1,600)
```
4.
```
   22
   613    (600)
    97    (100)
   452    (500)
   879    (900)
  + 30  + (000)
  2,071   (2,100)
```

5.
```
  1 2
  113    (100)
  251    (300)
  345    (300)
  355    (400)
+ 427  + (400)
―――――  ―――――
1,491   (1,500)
```

6.
```
  2 2
  546    (500)
  120    (100)
  309    (300)
  675    (700)
+ 481  + (500)
―――――  ―――――
2,131   (2,100)
```

7. $(100) + $(500) + $(200) = $(800)
   $127 + $475 + $225 = $827

8. (400) + (200) + (200) = (800)
   390 + 240 + 152 = 782 miles

Lesson Practice 18B

1.
```
  2 1
  264    (300)
   85    (100)
  624    (600)
+ 945  + (900)
―――――  ―――――
1,918   (1,900)
```

2.
```
  2 1
  172    (200)
  261    (300)
  527    (500)
+ 446  + (400)
―――――  ―――――
1,406   (1,400)
```

3.
```
  2 2
  933    (900)
   58    (100)
  361    (400)
  159    (200)
+ 542  + (500)
―――――  ―――――
2,053   (2,100)
```

4.
```
  2 1
  142    (100)
  206    (200)
  860    (900)
  462    (500)
+ 553  + (600)
―――――  ―――――
2,223   (2,300)
```

5.
```
  2 2
  321    (300)
   39    (000)
  686    (700)
  452    (500)
+ 152  + (200)
―――――  ―――――
1,650   (1,700)
```

6.
```
  2 2
  214    (200)
  596    (600)
  473    (500)
  527    (500)
+ 802  + (800)
―――――  ―――――
2,612   (2,600)
```

7. 208 + 316 + 365 = 889 gallons

8. 137 + 122 + 101 + 150 = 510 students

Lesson Practice 18C

1.
```
  1 2
  649    (600)
  536    (500)
   31    (000)
+ 224  + (200)
―――――  ―――――
1,440   (1,300)
```

2.
```
  1 2
  714    (700)
  746    (700)
  419    (400)
+ 652  + (700)
―――――  ―――――
2,531   (2,500)
```

3.
```
  2 2
  328    (300)
  530    (500)
  356    (400)
  432    (400)
+ 456  + (500)
―――――  ―――――
2,102   (2,100)
```

4.
```
 2 2
 2 1 9    (200)
 8 2 0    (800)
 4 7 9    (500)
 3 8 1    (400)
+1 6 1  + (200)
─────    ───────
2,060    (2,100)
```

5.
```
 1 2
 5 6 2    (600)
 5 2 0    (500)
    3 8    (000)
 6 5 9    (700)
+6 1 3  + (600)
─────    ───────
2,392    (2,400)
```

6.
```
 2 2
 2 4 7    (200)
 2 5 4    (300)
 4 1 6    (400)
 2 5 2    (300)
+5 4 7  + (500)
─────    ───────
1,716    (1,700)
```

7. 553 + 334 + 129 = 1,016 flakes

8. 302 + 148 + 447 = 897 cans

Systematic Review 18D

1.
```
 2 2
 2 9 6
 3 0 8
 7 4 2
 8 6 6
+3 1 4
─────
2,526
```

2.
```
 1 2
 5 2 1
    5 2
 6 2 4
 5 4 6
+   3 8
─────
1,781
```

3.
```
 2 1
 2 7 3
 9 5 1
 5 5 0
 3 3 9
+2 8 2
─────
2,395
```

4. 7,821; "seven thousand, eight hundred twenty-one"

5. 350,000

6. 2 − 0 = 2

7. 4 − 1 = 3

8. 6 − 2 = 4

9. 5 − 5 = 0

10. done

11. done

12. −

13. +

14. 2, 4, 6, 8, 10, 12, 14, 16, 18, 20

15. 51 + 29 + 19 = 99 guests

16. 68 + 89 + 132 = 289 flowers

Systematic Review 18E

1.
```
 2 1
 4 5 7
 5 6 1
 4 5 1
 6 3 1
+   2 9
─────
2,129
```

2.
```
 2 2
 8 7 3
 2 6 5
 3 1 4
 2 4 7
+9 3 6
─────
2,635
```

3.
```
 2 2
 8 0 4
 2 4 6
 5 3 1
 3 8 2
+6 3 9
─────
2,602
```

4. 81,246
   "eighty-one thousand,
   two hundred forty-six"
5. 652,693
6. $12 - 9 = 3$
7. $15 - 9 = 6$
8. $11 - 9 = 2$
9. $17 - 9 = 8$
10. +
11. –
12. –
13. +
14. 5, 10, 15, 20, 25, 30, 35, 40, 45, 50
15. $12 + 12 + 12 + 12 + 12 + 12 = 72$"
16. 3 nickels = 15¢; 4 dimes = 40¢
    15¢ + 40¢ = 55¢ or $0.55

Systematic Review 18F

1.
```
   1 1
   7 4 1
   4 0 5
   4 0 1
   5 8 6
 + 3 6 4
 ───────
   2,497
```

2.
```
   2 2
   9 5 2
   2 3 7
   1 7 1
   6 0 3
 + 4 5 9
 ───────
   2,422
```

3.
```
   2 1
   6 3 1
   8 3 4
   8 4 1
   2 7 2
 + 2 3 4
 ───────
   2,812
```

4. 637,531
   "six hundred thirty-seven thousand,
   five hundred thirty-one"
5. 45,727
6. $11 - 8 = 3$

7. $14 - 8 = 6$
8. $17 - 8 = 9$
9. $13 - 8 = 5$
10. –
11. –
12. +
13. –
14. 10, 20, 30, 40, 50, 60, 70, 80, 90, 100
15. $87 + 48 + 211 = 346$ pages
16. $16 - 7 = 9$ years

Lesson Practice 19A

1. done

2.
```
      1 1
   7,132    (7,000)
   5,333    (5,000)
 + 1,186  + (1,000)
 ────────  ────────
  13,651   (13,000)
```

3.
```
     1 1 1
   2,852    (3,000)
   4,263    (4,000)
 + 3,149  + (3,000)
 ────────  ────────
  10,264   (10,000)
```

4.
```
      1 1
   6,732    (7,000)
   3,152    (3,000)
 + 7,321  + (7,000)
 ────────  ────────
  17,205   (17,000)
```

5.
```
      1 1
   5,232    (5,000)
   7,111    (7,000)
 + 3,765  + (4,000)
 ────────  ────────
  16,108   (16,000)
```

6.
```
     1 1 1
   1,257    (1,000)
   6,463    (6,000)
 + 8,765  + (9,000)
 ────────  ────────
  16,485   (16,000)
```

7. $1,083 + 482 + 1,216 + 741 = 3,522$ miles
8. $3,187 + 408 + 1,443 + 1,837 = 6,875$ miles

Lesson Practice 19B

1.
```
  1 1 1
  2,817      (3,000)
  9,236      (9,000)
+ 3,680    + (4,000)
─────────────────────
 15,733     (16,000)
```

2.
```
  1
  5,740      (6,000)
  4,221      (4,000)
+ 3,321    + (3,000)
─────────────────────
 13,282     (13,000)
```

3.
```
  1 1 1
  3,213      (3,000)
  1,357      (1,000)
+ 2,798    + (3,000)
─────────────────────
  7,368      (7,000)
```

4.
```
  2 1 1
  1,476      (1,000)
    746      (1,000)
+ 9,813   + (10,000)
─────────────────────
 12,035     (12,000)
```

5.
```
  1 1
  2,741      (3,000)
  4,374      (4,000)
+ 3,354    + (3,000)
─────────────────────
 10,469     (10,000)
```

6.
```
  1 2 1
  4,123      (4,000)
  7,491      (7,000)
+   486    + (0,000)
─────────────────────
 12,100     (11,000)
```

7. 2,365 + 1,295 + 3,116 = 6,776 fish

8. 7,851 + 3,895 = 11,746 students

Lesson Practice 19C

1.
```
  1 2 1
  3,695      (4,000)
  3,175      (3,000)
+ 2,141    + (2,000)
─────────────────────
  9,011      (9,000)
```

2.
```
  1 1
  1,468      (1,000)
  6,012      (6,000)
+ 5,280    + (5,000)
─────────────────────
 12,760     (12,000)
```

3.
```
  2 1 1
  2,940      (3,000)
  4,278      (4,000)
+ 1,963    + (2,000)
─────────────────────
  9,181      (9,000)
```

4.
```
  1 1 2
  1,086      (1,000)
  3,608      (4,000)
+ 2,657    + (3,000)
─────────────────────
  7,351      (8,000)
```

5.
```
  1 1
  1,340      (1,000)
  1,765      (2,000)
+ 8,041    + (8,000)
─────────────────────
 11,146     (11,000)
```

6.
```
  1 1 1
  4,729      (5,000)
  7,761      (8,000)
+    35    + (0,000)
─────────────────────
 12,525     (13,000)
```

7. 8,716 + 6,658 = 15,374 people

8. $2,345 + $5,634 + $1,954 = $9,933

Systematic Review 19D

1.
```
  1 2 1
  2,475
  1,890
+   376
─────────
  4,741
```

2.
```
  1
  7,513
  9,025
+ 3,254
─────────
 19,792
```

3.
```
  1 1 1
  3,189
  1,422
+ 2,468
─────────
  7,079
```

4. 8 − 4 = 4

5. 10 − 6 = 4

6. 14 − 7 = 7

7. 10 − 9 = 1

8. 6 − 3 = 3

9. 10 − 3 = 7

10. $4 - 2 = 2$
11. $10 - 5 = 5$
12. $6 > 4$
13. $14 > 6$
14. $3 < 7$
15. $\$35 + \$42 + \$33 + \$45 = \$155$
16. $\$17 - \$8 = \$9$
17. $3,645 + 4,782 + 5,641 = 14,068$ earthworms
18. $5 + 5 + 5 = 15"$

Systematic Review 19E

1.
$$
\begin{array}{r}
\overset{1\ 1}{2,384} \\
4,123 \\
+6,335 \\
\hline 12,842
\end{array}
$$

2.
$$
\begin{array}{r}
\overset{1\ 2\ 1}{3,591} \\
2,367 \\
+8,459 \\
\hline 14,417
\end{array}
$$

3.
$$
\begin{array}{r}
\overset{1\ 1}{2,368} \\
4,152 \\
+6,314 \\
\hline 12,834
\end{array}
$$

4. $9 - 5 = 4$
5. $2 - 1 = 1$
6. $9 - 6 = 3$
7. $10 - 2 = 8$
8. $9 - 3 = 6$
9. $16 - 9 = 7$
10. $9 - 4 = 5$
11. $9 - 7 = 2$
12. $9 < 15$
13. $9 = 9$
14. $2 < 8$
15. $\$2.98 + \$3.75 = \$6.73$
16. $7 - 5 = 2$ dimes
    2 dimes = 20¢
17. $20 + 18 + 10 + 35 = 83$ tons
18. $35 + 35 + 35 + 35 = 140'$

Systematic Review 19F

1.
$$
\begin{array}{r}
\overset{1\ 1}{8,482} \\
4,621 \\
+5,351 \\
\hline 18,454
\end{array}
$$

2.
$$
\begin{array}{r}
\overset{1\ 1}{1,264} \\
7,632 \\
+1,953 \\
\hline 10,849
\end{array}
$$

3.
$$
\begin{array}{r}
\overset{1\ 1}{5,148} \\
2,633 \\
+4,186 \\
\hline 11,967
\end{array}
$$

4. $7 - 3 = 4$
5. $11 - 6 = 5$
6. $7 - 4 = 3$
7. $13 - 5 = 8$
8. $8 - 3 = 5$
9. $9 - 2 = 7$
10. $8 - 5 = 3$
11. $11 - 4 = 7$
12. $2 > 1$
13. $5 < 19$
14. $8 = 8$
15. $495 + 382 + 516 + 402 = 1,795$ miles
16. $25 + 35 + 25 + 35 = 120'$
17. 2, 4, 6, 8, 10, <u>12</u> mittens
18. $12 - 4 = 8$ mittens

Lesson Practice 20A

1. done

2.
$$
\begin{array}{r}
60 \\
-40 \\
\hline 20
\end{array}
\qquad
\begin{array}{r}
40 \\
+20 \\
\hline 60
\end{array}
$$

3.
$$
\begin{array}{r}
94 \\
-51 \\
\hline 43
\end{array}
\qquad
\begin{array}{r}
51 \\
+43 \\
\hline 94
\end{array}
$$

4.
$$
\begin{array}{r}
53 \\
-42 \\
\hline 11
\end{array}
\qquad
\begin{array}{r}
42 \\
+11 \\
\hline 53
\end{array}
$$

5.
```
  40    30
 -30   +10
  10    40
```

6.
```
 459   312
-312  +147
 147   459
```

7.
```
  44    20
 -20   +24
  24    44
```

8.
```
 924    13
 -13  +911
 911   924
```

9.
```
 506   302
-302  +204
 204   506
```

10.
```
  25    21
 -21   + 4
   4    25
```

11.
```
 841   620
-620  +221
 221   841
```

12.
```
 999   123
-123  +876
 876   999
```

13. 32 – 10 = 22 students
14. 44 – 11 = 33 rabbits

Lesson Practice 20B

1.
```
  35    24
 -24   +11
  11    35
```

2.
```
  26    13
 -13   +13
  13    26
```

3.
```
  50    20
 -20   +30
  30    50
```

4.
```
  83    12
 -12   +71
  71    83
```

5.
```
  49    47
 -47   + 2
   2    49
```

6.
```
 989   432
-432  +557
 557   989
```

7.
```
  46    30
 -30   +16
  16    46
```

8.
```
 554    21
 -21  +533
 533   554
```

9.
```
 300   100
-100  +200
 200   300
```

10.
```
  62    30
 -30   +32
  32    62
```

11.
```
 438   214
-214  +224
 224   438
```

12.
```
 397   175
-175  +222
 222   397
```

13. 68 – 25 = 43 cards
14. 48 – 23 = 25 gallons

Lesson Practice 20C

1.
```
  65    32
 -32   +33
  33    65
```

2.
```
  17    17
 -17   + 0
   0    17
```

3.
```
  52    21
 -21   +31
  31    52
```

4.
```
  20    10
 -10   +10
  10    20
```

5.
```
  75    32
 -32   +43
  43    75
```

6.
```
  188      23
-  23   +165
  165     188
```

7.
```
   69      31
 - 31    + 38
   38      69
```

8.
```
  561     260
 -260    +301
  301     561
```

9.
```
  645     435
 -435    +210
  210     645
```

10.
```
   85      44
 - 44    + 41
   41      85
```

11.
```
  225     100
 -100    +125
  125     225
```

12.
```
  538     421
 -421    +117
  117     538
```

13. $29 - 13 = 16$ children

14. $\$49 - \$21 = \$28$

7.
```
    1 1 1
   2,174
   7,418
 + 3,791
  13,383
```

8.
```
    1 1 1
   2,564
   6,408
 + 1,243
  10,215
```

9.
```
     1 2
    379
    511
    333
  + 468
  1,691
```

10. 124,971; "one hundred twenty-four thousand, nine hundred seventy-one"

11. 2, 4, 6, 8, 10, 12, 14, 16, 18, 20

12. 5, 10, 15, 20, 25, 30, 35, 40, 45, 50

13. 10, 20, 30, 40, 50, 60, 70, 80, 90, 100

14. $\$0.25 - \$0.14 = \$0.11$

15. $103 + 3,521 + 58 = 3,682$ miles

Systematic Review 20D

1.
```
   77      11
 - 11    + 66
   66      77
```

2.
```
   50      40
 - 40    + 10
   10      50
```

3.
```
   39      27
 - 27    + 12
   12      39
```

4.
```
  693     361
 -361    +332
  332     693
```

5.
```
  300     100
 -100    +200
  200     300
```

6.
```
  163      51
 - 51    +112
  112     163
```

Systematic Review 20E

1.
```
   35      22
 - 22    + 13
   13      35
```

2.
```
   49      31
 - 31    + 18
   18      49
```

3.
```
   28       5
 -  5    + 23
   23      28
```

4.
```
  633     510
 -510    +123
  123     633
```

5.
```
  790     690
 -690    +100
  100     790
```

6.
```
  561     550
 -550    + 11
   11     561
```

7.
```
  1 2
  1,890
  3,672
 +7,254
 12,816
```

8.
```
  1 2 1
  7,193
  4,685
 +1,492
 13,370
```

9.
```
  1 1
   922
   678
   253
  +112
  1,965
```

10. 205,918; "two hundred five thousand, nine hundred eighteen"
11. 5, 10, 15, 20, 25, 30, 35, 40, 45, 50
12. 2, 4, 6, 8, 10, 12, 14, 16, 18, 20
13. 10, 20, 30, 40, 50, 60, 70, 80, 90, 100
14. $0.49 – $0.23 = $0.26
15. 285 + 683 + 491 = 1,459 boxes

7.
```
  1 1 1
  2,467
  9,221
 +6,788
 18,476
```

8.
```
  1 1
  4,253
  4,112
 +1,255
  9,620
```

9.
```
  1 2
   238
   412
   633
  +459
  1,742
```

10. 23,684; "twenty-three thousand, six hundred eighty-four"
11. 10, 20, 30, 40, 50, 60, 70, 80, 90, 100
12. 5, 10, 15, 20, 25, 30, 35, 40, 45, 50
13. 2, 4, 6, 8, 10, 12, 14, 16, 18, 20
14. 12 + 12 + 12 = 36"
15. 67 – 13 = 54 cars

Systematic Review 20F

1.
```
 90    70
 70   +20
 20    90
```

2.
```
 51    10
-10   +41
 41    51
```

3.
```
 74    31
-31   +43
 43    74
```

4.
```
 139    125
-125   + 14
  14    139
```

5.
```
 999    347
-347   +652
 652    999
```

6.
```
 167     24
- 24   +143
 143    167
```

Lesson Practice 21A

1. done
2. :35
3. :40
4. :10
5. :48
6. :47

Lesson Practice 21B

1. :15
2. :25
3. :55
4. :50
5. :29
6. :16

Lesson Practice 21C
1. :05
2. :55
3. :40
4. :00
5. :08
6. :32

Systematic Review 21D
1. :10
2. :54
3.
```
  99      84
 -84     +15
  15      99
```
4.
```
 138      34
 -34    +104
 104     138
```
5.
```
 279     164
-164    +115
 115     279
```
6.
```
   1
$3.27
+4.16
$7.43
```
7.
```
   1
 2,384
+6,335
 8,719
```
8.
```
  1 1
  367
  591
 +419
 1,377
```
9. $4{,}568 + 3{,}851 + 2{,}001 = 10{,}420$ insects
10. $10 + 10 + 10 + 10 = 40'$ perimeter
    $40' - 3' = 37'$ of fence
11. $25 - 14 = 11$ years
12. $19 - 7 = 12$ points

Systematic Review 21E
1. :20
2. :49

3.
```
  83      52
 -52     +31
  31      83
```
4.
```
 873      61
 -61    +812
 812     873
```
5.
```
 362     151
-151    +211
 211     362
```
6.
```
   1
$ 9.28
+ 2.17
$11.45
```
7.
```
  4,123
 +7,460
 11,583
```
8.
```
  2 1
  459
  367
 +591
 1,417
```
9. $32 + 28 = 60$ years
10. $29 - 8 = 21$ years
11. $40 + 35 + 40 + 35 = 150'$
12. $144 + 120 = 264$ made
    $264 - 52 = 212$ eaten

Systematic Review 21F
1. :40
2. :17
3.
```
  19       7
 - 7     +12
  12      19
```
4.
```
 839      16
 -16    +823
 823     839
```
5.
```
 604     302
-302    +302
 302     604
```
6.
```
$5.36
+1.5 1
$6.87
```

7.
$$\begin{array}{r} \overset{1\ 1}{8,482} \\ +4,621 \\ \hline 13,103 \end{array}$$

8.
$$\begin{array}{r} \overset{1\ 1}{343} \\ 246 \\ +112 \\ \hline 701 \end{array}$$

9. 4 dimes = 40¢
3 nickels = 15¢
$0.40 + $0.15 = $0.55
$0.55 − $0.25 = $0.30

10. 6 in + 6 in + X in = 18 in
12 in + X in = 18 in; X = 6 in

11. 969 + 1,345 + 5,002 + 2,061 = 9,377 leaves

12. $25 − $10 = $15

Lesson Practice 22A

1. done

2.
$$\begin{array}{r} \overset{5}{6}\overset{}{4} \\ -27 \\ \hline 34 \end{array} \qquad \begin{array}{r} 27 \\ +34 \\ \hline 61 \end{array}$$

3.
$$\begin{array}{r} \overset{1}{2}\overset{1}{2} \\ -13 \\ \hline 9 \end{array} \qquad \begin{array}{r} 13 \\ +\ 9 \\ \hline 22 \end{array}$$

4.
$$\begin{array}{r} \overset{1}{2}\overset{1}{3} \\ -19 \\ \hline 4 \end{array} \qquad \begin{array}{r} 19 \\ +\ 4 \\ \hline 23 \end{array}$$

5.
$$\begin{array}{r} \overset{4}{5}\overset{1}{5} \\ -26 \\ \hline 29 \end{array} \qquad \begin{array}{r} 26 \\ +29 \\ \hline 55 \end{array}$$

6.
$$\begin{array}{r} \overset{2}{3}\overset{1}{2} \\ -14 \\ \hline 18 \end{array} \qquad \begin{array}{r} 14 \\ +18 \\ \hline 32 \end{array}$$

7.
$$\begin{array}{r} \overset{6}{7}\overset{1}{3} \\ -34 \\ \hline 39 \end{array} \qquad \begin{array}{r} 34 \\ +39 \\ \hline 73 \end{array}$$

8.
$$\begin{array}{r} \overset{5}{6}\overset{1}{8} \\ -29 \\ \hline 39 \end{array} \qquad \begin{array}{r} 29 \\ +39 \\ \hline 68 \end{array}$$

9.
$$\begin{array}{r} \overset{5}{6}\overset{1}{3} \\ -49 \\ \hline 14 \end{array} \qquad \begin{array}{r} 49 \\ +14 \\ \hline 63 \end{array}$$

10. $0.75 − $0.57 = $0.18

11. 43 − 28 = 15 barrettes

12. 72 − 65 = 7 peanuts

Lesson Practice 22B

1.
$$\begin{array}{r} \overset{4}{5}\overset{1}{7} \\ -29 \\ \hline 28 \end{array} \qquad \begin{array}{r} 29 \\ +28 \\ \hline 57 \end{array}$$

2.
$$\begin{array}{r} \overset{2}{3}\overset{1}{0} \\ -18 \\ \hline 12 \end{array} \qquad \begin{array}{r} 18 \\ +12 \\ \hline 30 \end{array}$$

3.
$$\begin{array}{r} \overset{5}{6}\overset{1}{5} \\ -47 \\ \hline 18 \end{array} \qquad \begin{array}{r} 47 \\ +18 \\ \hline 65 \end{array}$$

4.
$$\begin{array}{r} \overset{4}{5}\overset{1}{2} \\ -14 \\ \hline 38 \end{array} \qquad \begin{array}{r} 14 \\ +38 \\ \hline 52 \end{array}$$

5.
$$\begin{array}{r} \overset{7}{8}\overset{1}{5} \\ -49 \\ \hline 36 \end{array} \qquad \begin{array}{r} 49 \\ +36 \\ \hline 85 \end{array}$$

6.
$$\begin{array}{r} \overset{6}{7}\overset{1}{2} \\ -27 \\ \hline 45 \end{array} \qquad \begin{array}{r} 27 \\ +45 \\ \hline 72 \end{array}$$

7.
$$\begin{array}{r} \overset{4}{5}\overset{1}{3} \\ -18 \\ \hline 35 \end{array} \qquad \begin{array}{r} 18 \\ +35 \\ \hline 53 \end{array}$$

8.
$$\begin{array}{r} \overset{2}{3}\overset{1}{4} \\ -16 \\ \hline 15 \end{array} \qquad \begin{array}{r} 16 \\ +15 \\ \hline 31 \end{array}$$

9.
$$
\begin{array}{cc}
\overset{7}{\cancel{8}}\overset{1}{2} & 53 \\
-5\,3 & +29 \\
\hline
2\,9 & 82
\end{array}
$$

10. $43 - 24 = 19$ pages

11. $61 - 36 = 25$ times

12. $\$0.50 - \$0.23 = \$0.27$

9.
$$
\begin{array}{cc}
\overset{6}{\cancel{7}}\overset{1}{2} & 38 \\
-3\,8 & +34 \\
\hline
3\,4 & 72
\end{array}
$$

10. done

11. $48 - X = 19;\ X = 29$ years

12. $63 - X = 39;\ X = 24$ apples

Lesson Practice 22C

1.
$$
\begin{array}{cc}
\overset{4}{\cancel{5}}\overset{1}{2} & 25 \\
-2\,5 & +27 \\
\hline
2\,7 & 52
\end{array}
$$

2.
$$
\begin{array}{cc}
\overset{6}{\cancel{7}}\overset{1}{1} & 3 \\
-\ \ 3 & +68 \\
\hline
6\,8 & 71
\end{array}
$$

3.
$$
\begin{array}{cc}
\overset{2}{\cancel{3}}\overset{1}{4} & 15 \\
-1\,5 & +19 \\
\hline
1\,9 & 34
\end{array}
$$

4.
$$
\begin{array}{cc}
\overset{7}{\cancel{8}}\overset{1}{7} & 8 \\
-\ \ 8 & +79 \\
\hline
7\,9 & 87
\end{array}
$$

5.
$$
\begin{array}{cc}
\overset{5}{\cancel{6}}\overset{1}{2} & 27 \\
-2\,7 & +35 \\
\hline
3\,5 & 62
\end{array}
$$

6.
$$
\begin{array}{cc}
\overset{1}{\cancel{2}}\overset{1}{3} & 14 \\
-1\,4 & +\ 9 \\
\hline
9 & 23
\end{array}
$$

7.
$$
\begin{array}{cc}
\overset{3}{\cancel{4}}\overset{1}{5} & 26 \\
-2\,6 & +19 \\
\hline
1\,9 & 45
\end{array}
$$

8.
$$
\begin{array}{cc}
\overset{2}{\cancel{3}}\overset{1}{1} & 29 \\
-2\,9 & +\ 2 \\
\hline
2 & 31
\end{array}
$$

Systematic Review 22D

1.
$$
\begin{array}{cc}
\overset{6}{\cancel{7}}\overset{1}{1} & 43 \\
-4\,3 & +28 \\
\hline
2\,8 & 71
\end{array}
$$

2.
$$
\begin{array}{cc}
\overset{8}{\cancel{9}}\overset{1}{8} & 89 \\
-8\,9 & +\ 9 \\
\hline
9 & 98
\end{array}
$$

3.
$$
\begin{array}{cc}
\overset{2}{\cancel{3}}\overset{1}{5} & 17 \\
-1\,7 & +18 \\
\hline
1\,8 & 35
\end{array}
$$

4.
$$
\begin{array}{cc}
85 & 45 \\
-4\,5 & +40 \\
\hline
4\,0 & 85
\end{array}
$$

5.
$$
\begin{array}{cc}
429 & 311 \\
-3\,1\,1 & +1\,18 \\
\hline
1\,1\,8 & 429
\end{array}
$$

6.
$$
\begin{array}{cc}
148 & 26 \\
-\ \ 26 & +122 \\
\hline
1\,2\,2 & 148
\end{array}
$$

7.
$$
\begin{array}{r}
\overset{1}{\$7.33} \\
+1.83 \\
\hline
\$9.16
\end{array}
$$

8.
$$
\begin{array}{r}
\overset{1}{5{,}263} \\
+7{,}554 \\
\hline
12{,}817
\end{array}
$$

9.
$$
\begin{array}{r}
\overset{1\ 1}{892} \\
413 \\
+476 \\
\hline
1{,}781
\end{array}
$$

10. $:13$

11. : 45

12. 16 + 21 + 14 = 51

    51 − 25 = 26 marbles

Systematic Review 22E

1.
$$
\begin{array}{cc}
\overset{8}{\cancel{9}}\,{}^{1}2 & 78 \\
-\,7\,8 & +14 \\
\hline
1\,4 & 92 \\
\end{array}
$$

2.
$$
\begin{array}{cc}
\overset{4}{\cancel{5}}\,{}^{1}3 & 45 \\
-\,4\,5 & +\,8 \\
\hline
8 & 53 \\
\end{array}
$$

3.
$$
\begin{array}{cc}
\overset{1}{\cancel{2}}\,{}^{1}3 & 14 \\
-\,1\,4 & +\,9 \\
\hline
9 & 23 \\
\end{array}
$$

4.
$$
\begin{array}{cc}
75 & 50 \\
-\,50 & +25 \\
\hline
25 & 75 \\
\end{array}
$$

5.
$$
\begin{array}{cc}
7\,4\,3 & 12 \\
-\ \ 1\,2 & +731 \\
\hline
731 & 743 \\
\end{array}
$$

6.
$$
\begin{array}{cc}
5\,1\,4 & 512 \\
-\ \ \ 2 & +\ \ 2 \\
\hline
512 & 514 \\
\end{array}
$$

7.
$$
\begin{array}{r}
\overset{1}{\ }\ \\
\$3.55 \\
+2.52 \\
\hline
\$6.07 \\
\end{array}
$$

8.
$$
\begin{array}{r}
9,435 \\
+4,252 \\
\hline
13,687 \\
\end{array}
$$

9.
$$
\begin{array}{r}
{}^{2\,1}\ \ \ \\
573 \\
882 \\
235 \\
+762 \\
\hline
2,452 \\
\end{array}
$$

10. : 30

11. : 53

12. 19 + 21 = 40

    52 − 40 = 12 pages

Systematic Review 22F

1.
$$
\begin{array}{cc}
\overset{3}{\cancel{4}}\,{}^{1}4 & 28 \\
-\,2\,8 & +16 \\
\hline
1\,6 & 44 \\
\end{array}
$$

2.
$$
\begin{array}{cc}
\overset{8}{\cancel{9}}\,{}^{1}3 & 16 \\
-\,1\,6 & +77 \\
\hline
7\,7 & 93 \\
\end{array}
$$

3.
$$
\begin{array}{cc}
\overset{7}{\cancel{8}}\,{}^{1}4 & 27 \\
-\,2\,7 & +57 \\
\hline
5\,7 & 84 \\
\end{array}
$$

4.
$$
\begin{array}{cc}
17 & 10 \\
-\,10 & +\ 7 \\
\hline
7 & 17 \\
\end{array}
$$

5.
$$
\begin{array}{cc}
913 & 112 \\
-\,112 & +801 \\
\hline
801 & 913 \\
\end{array}
$$

6.
$$
\begin{array}{cc}
307 & 205 \\
-\,205 & +102 \\
\hline
102 & 307 \\
\end{array}
$$

7.
$$
\begin{array}{r}
{}^{1}\ \ \ \\
\$\ 6.24 \\
+\ 5.48 \\
\hline
\$11.72 \\
\end{array}
$$

8.
$$
\begin{array}{r}
{}^{1\ 1\ 1}\ \ \\
3,548 \\
7,624 \\
+3,642 \\
\hline
14,814 \\
\end{array}
$$

9.
$$
\begin{array}{r}
{}^{2}\ \ \ \\
573 \\
882 \\
+762 \\
\hline
2,217 \\
\end{array}
$$

10. : 02

11. : 15

12. Bill : 134 + 48 + 2 = 184 animals

    Tom : 100 + 22 = 122 animals

    184 − 122 = 62 more

Lesson Practice 23A
1. done
2. 7:00
3. 11:00
4. 1:00

Lesson Practice 23B
1. 8:00
2. 2:00
3. 12:00
4. 4:00

Lesson Practice 23C
1. done
2. 5:45
3. 3:40
4. 12:25

Systematic Review 23D
1. 5:50
2. 2:40
3. 7:09
4. 4:17
5. 40 − 38 = 2; 38 + 2 = 40
6. 76 − 18 = 58; 18 + 58 = 76
7. 62 − 57 = 5; 57 + 5 = 62
8. 73 − 14 = 59; 14 + 59 = 73
9. $0.40 + $0.35 + $0.08 = $0.83
10. 1,342 + 1.342 + 1,342 + 1,342 = 5,368'

Systematic Review 23E
1. 6:00
2. 8:05
3. 1:21
4. 9:57
5. 73 − 15 = 58; 15 + 58 = 73
6. 88 − 44 = 44; 44 + 44 = 88
7. 76 − 28 = 48; 28 + 48 = 76
8. 928 − 605 = 323; 605 + 323 = 928
9. 12+12+12+12+12+12 = 72"
10. 250 + 75 = 325 soldiers

Systematic Review 23F
1. 10:15
2. 3:30
3. 12:08
4. 7:45
5. 54 − 45 = 9; 45 + 9 = 54
6. 39 − 20 = 19; 20 + 19 = 39
7. 40 − 35 = 5; 35 + 5 = 40
8. 779 − 314 = 465; 314 + 465 = 779
9. $4.52 + $2.91 = $7.43
10. 35 − 26 = 9 birds

Lesson Practice 24A

1. done
2. done

3.
$$
\begin{array}{r} 2\overset{1}{\cancel{2}}\overset{\overset{1}{4}}{\cancel{3}} \\ -\ 87 \\ \hline 136 \end{array}
\qquad
\begin{array}{r} 87 \\ +136 \\ \hline 223 \end{array}
$$

4.
$$
\begin{array}{r} 7\overset{6}{\cancel{3}}\overset{\overset{2}{4}}{} \\ -\ 36 \\ \hline 698 \end{array}
\qquad
\begin{array}{r} 36 \\ +698 \\ \hline 734 \end{array}
$$

5.
$$
\begin{array}{r} 5\overset{4}{\cancel{2}}\overset{\overset{1}{3}}{\cancel{3}} \\ -138 \\ \hline 385 \end{array}
\qquad
\begin{array}{r} 138 \\ +385 \\ \hline 523 \end{array}
$$

6.
$$
\begin{array}{r} 4\overset{3}{\cancel{0}}\overset{9}{\cancel{0}} \\ -399 \\ \hline 1 \end{array}
\qquad
\begin{array}{r} 399 \\ +\ \ 1 \\ \hline 400 \end{array}
$$

7.
$$
\begin{array}{r} 3\overset{2}{\cancel{2}}\overset{\overset{1}{4}}{\cancel{4}} \\ -\ 39 \\ \hline 282 \end{array}
\qquad
\begin{array}{r} 39 \\ +282 \\ \hline 321 \end{array}
$$

8.
$$
\begin{array}{r} 459 \\ -241 \\ \hline 218 \end{array}
\qquad
\begin{array}{r} 241 \\ +218 \\ \hline 459 \end{array}
$$

9.
$$
\begin{array}{r} 6\overset{5}{\cancel{1}}\overset{6}{\cancel{4}} \\ -196 \\ \hline 418 \end{array}
\qquad
\begin{array}{r} 196 \\ +418 \\ \hline 614 \end{array}
$$

10. $235 - 167 = 68$ pages
11. $300 - 245 = 55$ pennies
12. $580 - 425 = 155'$

Lesson Practice 24B

1.
$$
\begin{array}{r} 7\overset{6}{\cancel{1}}\overset{6}{\cancel{6}} \\ -\ 58 \\ \hline 658 \end{array}
\qquad
\begin{array}{r} 58 \\ +658 \\ \hline 716 \end{array}
$$

2.
$$
\begin{array}{r} 6\overset{5}{\cancel{6}}8 \\ -284 \\ \hline 384 \end{array}
\qquad
\begin{array}{r} 284 \\ +384 \\ \hline 668 \end{array}
$$

3.
$$
\begin{array}{r} 163 \\ -\ 72 \\ \hline 91 \end{array}
\qquad
\begin{array}{r} \overset{1}{7}2 \\ +\ 91 \\ \hline 163 \end{array}
$$

4.
$$
\begin{array}{r} 6\overset{5}{\cancel{4}}\overset{\overset{3}{4}}{} \\ -\ 49 \\ \hline 592 \end{array}
\qquad
\begin{array}{r} 49 \\ +592 \\ \hline 641 \end{array}
$$

5.
$$
\begin{array}{r} 3\overset{6}{\cancel{7}}2 \\ -106 \\ \hline 266 \end{array}
\qquad
\begin{array}{r} 106 \\ +266 \\ \hline 372 \end{array}
$$

6.
$$
\begin{array}{r} 9\overset{8}{\cancel{8}}\overset{7}{\cancel{7}} \\ -789 \\ \hline 198 \end{array}
\qquad
\begin{array}{r} 789 \\ +198 \\ \hline 987 \end{array}
$$

7.
$$
\begin{array}{r} 7\overset{6}{\cancel{0}}5 \\ -\ 60 \\ \hline 645 \end{array}
\qquad
\begin{array}{r} 60 \\ +645 \\ \hline 705 \end{array}
$$

8.
$$
\begin{array}{r} 5\overset{4}{\cancel{0}}0 \\ -250 \\ \hline 250 \end{array}
\qquad
\begin{array}{r} 250 \\ +250 \\ \hline 500 \end{array}
$$

9.
$$
\begin{array}{r} 4\overset{6}{\cancel{7}}\overset{\overset{1}{7}}{} \\ -106 \\ \hline 365 \end{array}
\qquad
\begin{array}{r} 106 \\ +365 \\ \hline 471 \end{array}
$$

10. $\$275 - \$116 = \$159$
11. $553 - 378 = 175$ acorns
12. $325 - 185 = 140$ miles

Lesson Practice 24C

1.
$$
\begin{array}{r} 5\overset{7}{\cancel{8}}\overset{\overset{1}{3}}{\cancel{3}} \\ -\ 64 \\ \hline 519 \end{array}
\qquad
\begin{array}{r} 64 \\ +519 \\ \hline 583 \end{array}
$$

2.
$$
\begin{array}{r} \overset{1}{2}\overset{}{0}3 \\ -192 \\ \hline 11 \end{array}
\qquad
\begin{array}{r} \overset{1}{1}92 \\ +\ 11 \\ \hline 203 \end{array}
$$

3.
$$
\begin{array}{r} 2\overset{1}{\cancel{0}}\overset{9}{\cancel{0}} \\ -\ 98 \\ \hline 102 \end{array}
\qquad
\begin{array}{r} 98 \\ +102 \\ \hline 200 \end{array}
$$

4.
$$
\begin{array}{r}
2 \\
3\,\overset{9}{\cancel{1}}9 \\
-\ \ 30 \\
\hline
289
\end{array}
\qquad
\begin{array}{r}
1 \\
30 \\
+289 \\
\hline
319
\end{array}
$$

5.
$$
\begin{array}{r}
5 \\
6\,\overset{}{\cancel{3}} \\
-\ 24 \\
\hline
39
\end{array}
\qquad
\begin{array}{r}
1 \\
24 \\
+39 \\
\hline
63
\end{array}
$$

5.
$$
\begin{array}{r}
2 \\
6\,\overset{}{\cancel{3}}4 \\
-429 \\
\hline
202
\end{array}
\qquad
\begin{array}{r}
1 \\
429 \\
+202 \\
\hline
631
\end{array}
$$

6.
$$
\begin{array}{r}
3 \\
4\,\overset{}{\cancel{1}} \\
-\ 35 \\
\hline
6
\end{array}
\qquad
\begin{array}{r}
1 \\
35 \\
+\ 6 \\
\hline
41
\end{array}
$$

6.
$$
\begin{array}{r}
3 \\
4\,\overset{9}{\cancel{1}}9 \\
-138 \\
\hline
281
\end{array}
\qquad
\begin{array}{r}
1 \\
138 \\
+281 \\
\hline
419
\end{array}
$$

7. 4:35

8. 11:12

7.
$$
\begin{array}{r}
9 \\
\overset{}{\cancel{1}}\overset{}{\cancel{0}}3 \\
-\ 25 \\
\hline
78
\end{array}
\qquad
\begin{array}{r}
11 \\
25 \\
+\ 78 \\
\hline
103
\end{array}
$$

9.
$$
\begin{array}{r}
436 \\
+122 \\
\hline
558
\end{array}
$$

10.
$$
\begin{array}{r}
11 \\
298 \\
+539 \\
\hline
837
\end{array}
$$

8.
$$
\begin{array}{r}
4 \\
5\,\overset{}{\cancel{7}}2 \\
-390 \\
\hline
182
\end{array}
\qquad
\begin{array}{r}
1 \\
390 \\
+182 \\
\hline
572
\end{array}
$$

11.
$$
\begin{array}{r}
111 \\
2{,}999 \\
+3{,}111 \\
\hline
6{,}110
\end{array}
$$

9.
$$
\begin{array}{r}
2\ \ 2 \\
\overset{}{\cancel{3}}\,\overset{}{\cancel{3}}3 \\
-144 \\
\hline
189
\end{array}
\qquad
\begin{array}{r}
11 \\
144 \\
+189 \\
\hline
333
\end{array}
$$

12. 175 – 98 = 77 fireflies

13. 212 + 362 = 574 miles

14. $215 + $134 = $349

$400 – $349 = $51

10. 875 – 80 = 795 people

11. 200 – 115 = 85 cards

12. 375 – 18 = 357 yards

Systematic Review 24D

1.
$$
\begin{array}{r}
9 \\
\overset{}{\cancel{1}}\overset{}{\cancel{0}}0 \\
-\ 75 \\
\hline
25
\end{array}
\qquad
\begin{array}{r}
11 \\
75 \\
+\ 25 \\
\hline
100
\end{array}
$$

2.
$$
\begin{array}{r}
8 \\
\overset{}{\cancel{9}}08 \\
-291 \\
\hline
617
\end{array}
\qquad
\begin{array}{r}
1 \\
291 \\
+617 \\
\hline
908
\end{array}
$$

3.
$$
\begin{array}{r}
4 \\
2\,\overset{}{\cancel{5}}6 \\
-138 \\
\hline
118
\end{array}
\qquad
\begin{array}{r}
1 \\
138 \\
+118 \\
\hline
256
\end{array}
$$

4.
$$
\begin{array}{r}
199 \\
-\ 82 \\
\hline
117
\end{array}
\qquad
\begin{array}{r}
82 \\
+117 \\
\hline
199
\end{array}
$$

Systematic Review 24E

1.
$$
\begin{array}{r}
4 \\
1\,\overset{}{\cancel{5}}2 \\
-\ 87 \\
\hline
65
\end{array}
\qquad
\begin{array}{r}
11 \\
87 \\
+\ 65 \\
\hline
152
\end{array}
$$

2.
$$
\begin{array}{r}
1 \\
3\,\overset{}{\cancel{2}}0 \\
-118 \\
\hline
202
\end{array}
\qquad
\begin{array}{r}
1 \\
118 \\
+202 \\
\hline
320
\end{array}
$$

3.
$$
\begin{array}{r}
803 \\
-300 \\
\hline
503
\end{array}
\qquad
\begin{array}{r}
300 \\
+503 \\
\hline
803
\end{array}
$$

4.
$$
\begin{array}{r}
8 \\
1\,\overset{}{\cancel{9}}3 \\
-\ 67 \\
\hline
126
\end{array}
\qquad
\begin{array}{r}
1 \\
67 \\
+126 \\
\hline
193
\end{array}
$$

5.
$$\begin{array}{r} \overset{3}{\cancel{4}}\overset{1}{\cancel{3}} \\ -1\,7 \\ \hline 2\,6 \end{array} \qquad \begin{array}{r} 17 \\ +2\,6 \\ \hline 4\,3 \end{array}$$

6.
$$\begin{array}{r} \overset{5}{\cancel{6}}\overset{1}{\cancel{4}} \\ -3\,8 \\ \hline 2\,6 \end{array} \qquad \begin{array}{r} 38 \\ +2\,6 \\ \hline 6\,4 \end{array}$$

7. 2:05

8. 4:48

9.
$$\begin{array}{r} 2\,1\,2 \\ +3\,6\,2 \\ \hline 5\,7\,4 \end{array}$$

10.
$$\begin{array}{r} \overset{1}{} \\ 3\,5\,6 \\ +4\,8\,1 \\ \hline 8\,3\,7 \end{array}$$

11.
$$\begin{array}{r} \overset{1}{} \\ 1,2\,7\,6 \\ +7,3\,9\,1 \\ \hline 8,6\,6\,7 \end{array}$$

12. $321 - 215 = 106$ pages

13. $665 + 133 = 798$ animals

14. $314 + 219 = 533$
     $658 - 533 = 125$ cones

Systematic Review 24F

1.
$$\begin{array}{r} \overset{9}{\cancel{1}}\overset{1}{\cancel{0}}\,0 \\ -4\,7 \\ \hline 5\,3 \end{array} \qquad \begin{array}{r} 47 \\ +5\,3 \\ \hline 1\,0\,0 \end{array}$$

2.
$$\begin{array}{r} \overset{3}{1\,4}\overset{1}{\cancel{1}} \\ -1\,1\,3 \\ \hline 2\,8 \end{array} \qquad \begin{array}{r} 113 \\ +2\,8 \\ \hline 1\,4\,1 \end{array}$$

3.
$$\begin{array}{r} \overset{1}{\cancel{2}}\overset{4}{\cancel{2}}\,0 \\ -1\,6\,4 \\ \hline 5\,6 \end{array} \qquad \begin{array}{r} 164 \\ +5\,6 \\ \hline 2\,2\,0 \end{array}$$

4.
$$\begin{array}{r} \overset{4}{1\,5}\overset{1}{\cancel{0}} \\ -9\,8 \\ \hline 5\,2 \end{array} \qquad \begin{array}{r} 98 \\ +5\,2 \\ \hline 1\,5\,0 \end{array}$$

5.
$$\begin{array}{r} 70 \\ -2\,0 \\ \hline 5\,0 \end{array} \qquad \begin{array}{r} 20 \\ +5\,0 \\ \hline 7\,0 \end{array}$$

6.
$$\begin{array}{r} \overset{2}{\cancel{3}}\overset{1}{\cancel{2}} \\ -1\,8 \\ \hline 1\,4 \end{array} \qquad \begin{array}{r} 18 \\ +1\,4 \\ \hline 3\,2 \end{array}$$

7. 5:22

8. 10:36

9.
$$\begin{array}{r} 4\,0\,0 \\ +3\,7\,6 \\ \hline 7\,7\,6 \end{array}$$

10.
$$\begin{array}{r} \overset{1\,1}{} \\ 2\,8\,8 \\ +1\,3\,7 \\ \hline 4\,2\,5 \end{array}$$

11.
$$\begin{array}{r} \overset{1}{} \\ 2,1\,4\,6 \\ +1,4\,0\,8 \\ \hline 3,5\,5\,4 \end{array}$$

12. $581 - 499 = 82$ people

13. $\$126 + \$132 = \$258$

14. $45 + 11 + 5 + 8 = 69$
     $100 - 69 = 31$ dogs

Lesson Practice 25A

1. February

2. second

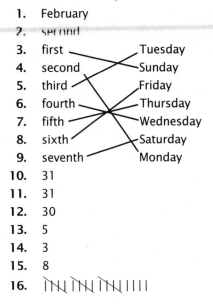

3. first — Tuesday
4. second — Sunday
5. third — Friday
6. fourth — Thursday
7. fifth — Wednesday
8. sixth — Saturday
9. seventh — Monday

10. 31

11. 31

12. 30

13. 5

14. 3

15. 8

16. |||| |||| |||| ||||

17. 卌 卌 卌 卌 卌 卌
18. 卌 卌 卌 卌 卌 IIII
19. 卌 卌 卌 卌 卌 卌 卌
20. sixth

Slash lines on tally marks
may slant in either direction

Lesson Practice 25B

1. Wednesday
2. seventh
3. October
4. first
5. second
6. third
7. fourth
8. fifth
9. sixth

first → January
second → April
third → May
fourth → February
fifth → June
sixth → March

10. 28 or 29
11. 31
12. 30
13. 11
14. 6
15. 15
16. 卌 卌
17. 卌 卌 III
18. 卌 II
19. 17 cars
20. Sunday

Lesson Practice 25C

1. April
2. third
3. first
4. seventh
5. eighth
6. ninth
7. tenth
8. eleventh
9. twelfth

seventh → November
eighth → August
ninth → September
tenth → October
eleventh → July
twelfth → December

10. 30
11. 31
12. 31
13. 12
14. 8
15. 14
16. 卌 卌 卌 卌 卌
17. 卌 IIII
18. IIII
19. Thursday
20. Valentine's Day

Systematic Review 25D

1. January
2. fourth
3. sixth
4. 31

5.
$$\begin{array}{r} 3\overset{4}{\cancel{5}}{}^{1}5 \\ -\ 26 \\ \hline 329 \end{array}$$

6.
$$\begin{array}{r} \overset{5}{\cancel{6}}{}^{1}18 \\ -\ 224 \\ \hline 394 \end{array}$$

7.
$$\begin{array}{r} \overset{3}{\cancel{4}}{}^{1}19 \\ -\ 357 \\ \hline 62 \end{array}$$

8.
$$\begin{array}{r} {}^{1}\ \ {}^{1}\\ 3,629 \\ +2,428 \\ \hline 6,057 \end{array}$$

9. 7:15
10. 4:55
11. 12 + 8 = 20 times
12. 5 + 17 + 14 = 36 beans
    卌 卌 卌 卌 卌 卌 卌 I
13. 131 - 47 = 84 days
14. 11 + 11 + 11 + 11 = 44'

Systematic Review 25E

1. March
2. Friday
3. September
4. 30

5.
$$\begin{array}{r} \overset{1}{\cancel{2}}\overset{1}{3}4 \\ -52 \\ \hline 182 \end{array}$$

6.
$$\begin{array}{r} 5\overset{8}{\cancel{9}}0 \\ -418 \\ \hline 172 \end{array}$$

7.
$$\begin{array}{r} \overset{8}{\cancel{9}}\overset{0}{\cancel{1}}0 \\ -75 \\ \hline 835 \end{array}$$

8.
$$\begin{array}{r} \overset{1}{1},\overset{1}{8}31 \\ +5,439 \\ \hline 7,270 \end{array}$$

9. 3:00
10. 6:30
11. 12 times
12. Answers will vary.
13. $3,451+2,163+1,999 = 7,613$ miles
14. August

Systematic Review 25F

1. May
2. November
3. sixth
4. 31

5.
$$\begin{array}{r} \overset{4}{\cancel{5}}07 \\ -41 \\ \hline 466 \end{array}$$

6.
$$\begin{array}{r} 1\overset{7}{\cancel{8}}2 \\ -146 \\ \hline 36 \end{array}$$

7.
$$\begin{array}{r} \overset{8}{\cancel{9}}\overset{9}{\cancel{0}}0 \\ -25 \\ \hline 875 \end{array}$$

8.
$$\begin{array}{r} \overset{1}{7},147 \\ +8,713 \\ \hline 15,860 \end{array}$$

9. 5:22
10. 10:05
11. $19 < 21$
    David read more.
12. seventh
13. $110 - 35 = 75$ years
14. 6 in + 10 in + X in = 29 in
    16 in + X in = 29 in; X = 13 in

Lesson Practice 26A

1. done
2. done

3.
$$\begin{array}{r} 5,\overset{2}{\cancel{3}}\overset{2}{\cancel{3}}3 \\ -1,186 \\ \hline 4,147 \end{array} \qquad \begin{array}{r} \overset{1}{1},\overset{1}{1}86 \\ +4,147 \\ \hline 5,333 \end{array}$$

4.
$$\begin{array}{r} 4,\overset{3}{\cancel{2}}\overset{7}{\cancel{8}}4 \\ -955 \\ \hline 3,329 \end{array} \qquad \begin{array}{r} \overset{1}{9}\overset{1}{5}5 \\ +3,329 \\ \hline 4,284 \end{array}$$

5.
$$\begin{array}{r} 8,2\overset{5}{\cancel{6}}3 \\ -3,149 \\ \hline 5,114 \end{array} \qquad \begin{array}{r} 3,\overset{1}{1}49 \\ +5,114 \\ \hline 8,263 \end{array}$$

6.
$$\begin{array}{r} 3,\overset{2}{\cancel{2}}\overset{4}{\cancel{6}}4 \\ -2,582 \\ \hline 682 \end{array} \qquad \begin{array}{r} \overset{1}{6}\overset{1}{8}2 \\ +2,582 \\ \hline 3,264 \end{array}$$

7.
$$\begin{array}{r} 7,\overset{6}{\cancel{2}}\overset{4}{\cancel{4}}\overset{3}{\cancel{1}} \\ -378 \\ \hline 6,863 \end{array} \qquad \begin{array}{r} \overset{1}{3}\overset{1}{7}\overset{1}{8} \\ +6,863 \\ \hline 7,241 \end{array}$$

8.
$$\begin{array}{r} 6,\overset{5}{\cancel{0}}\overset{9}{\cancel{0}}\overset{9}{\cancel{0}} \\ -5,139 \\ \hline 861 \end{array} \qquad \begin{array}{r} \overset{1}{5},\overset{1}{1}\overset{1}{3}9 \\ +861 \\ \hline 6,000 \end{array}$$

9.
$$\begin{array}{r} 6,\overset{6}{\cancel{7}}32 \\ -3,152 \\ \hline 3,580 \end{array} \qquad \begin{array}{r} 3,\overset{1}{5}80 \\ +3,152 \\ \hline 6,732 \end{array}$$

10. $1,465 - 906 = 559$ ants

11. $2,375 - 1,490 = 885$ miles
12. $1,760 - 1,588 = 172$ years

Lesson Practice 26B

1. $\begin{array}{r} \overset{4}{5},089 \\ - \ \ 632 \\ \hline 4,457 \end{array}$ $\begin{array}{r} \overset{1}{4},457 \\ + \ \ 632 \\ \hline 5,089 \end{array}$

2. $\begin{array}{r} \overset{6}{7},\overset{1}{3}2\overset{}{4} \\ -2,514 \\ \hline 4,807 \end{array}$ $\begin{array}{r} \overset{1}{2},\overset{1}{5}14 \\ +4,807 \\ \hline 7,321 \end{array}$

3. $\begin{array}{r} \overset{8}{9},\overset{9}{0}\overset{9}{0}0 \\ -1,287 \\ \hline 7,713 \end{array}$ $\begin{array}{r} \overset{1}{1},\overset{1}{2}87 \\ +7,713 \\ \hline 9,000 \end{array}$

4. $\begin{array}{r} \overset{6}{7},\overset{0}{1}\overset{0}{1}\overset{}{1} \\ - \ \ 232 \\ \hline 6,879 \end{array}$ $\begin{array}{r} \overset{1}{2}32 \\ +6,879 \\ \hline 7,111 \end{array}$

5. $\begin{array}{r} \overset{4}{5},\overset{2}{3}\overset{5}{6}1 \\ -3,765 \\ \hline 1,596 \end{array}$ $\begin{array}{r} \overset{1}{3},\overset{1}{7}65 \\ +1,596 \\ \hline 5,361 \end{array}$

6. $\begin{array}{r} 7,2\overset{0}{1}\overset{}{4} \\ -1,108 \\ \hline 6,106 \end{array}$ $\begin{array}{r} \overset{1}{1},108 \\ +6,106 \\ \hline 7,214 \end{array}$

7. $\begin{array}{r} 6,\overset{3}{4}\overset{9}{0}3 \\ - \ \ 257 \\ \hline 6,146 \end{array}$ $\begin{array}{r} \overset{1}{2}\overset{1}{5}7 \\ +6,146 \\ \hline 6,403 \end{array}$

8. $\begin{array}{r} 8,\overset{6}{7}65 \\ -3,085 \\ \hline 5,680 \end{array}$ $\begin{array}{r} \overset{1}{3},085 \\ +5,680 \\ \hline 8,765 \end{array}$

9. $\begin{array}{r} 4,987 \\ -3,732 \\ \hline 1,255 \end{array}$ $\begin{array}{r} 3,732 \\ +1,255 \\ \hline 4,987 \end{array}$

10. $\$1,579 - \$890 = \$689$
11. $3,451 - 2,999 = 452$ miles
12. $1,976 - 1,917 = 59$ years

Lesson Practice 26C

1. $\begin{array}{r} 1,6\overset{4}{5}\overset{}{0} \\ - \ \ 943 \\ \hline 707 \end{array}$ $\begin{array}{r} \overset{1}{9}\overset{1}{4}3 \\ + \ \ 707 \\ \hline 1,650 \end{array}$

2. $\begin{array}{r} \overset{7}{8},\overset{1}{2}\overset{9}{0}0 \\ -2,817 \\ \hline 5,383 \end{array}$ $\begin{array}{r} \overset{1}{2},\overset{1}{8}\overset{1}{1}7 \\ +5,383 \\ \hline 8,200 \end{array}$

3. $\begin{array}{r} \overset{4}{5},\overset{1}{2}21 \\ -4,740 \\ \hline 481 \end{array}$ $\begin{array}{r} \overset{1}{4},\overset{1}{7}40 \\ + \ \ 481 \\ \hline 5,221 \end{array}$

4. $\begin{array}{r} 8,\overset{6}{7}\overset{5}{6}5 \\ - \ \ 678 \\ \hline 8,087 \end{array}$ $\begin{array}{r} \overset{1}{6}\overset{1}{7}8 \\ +8,087 \\ \hline 8,765 \end{array}$

5. $\begin{array}{r} 2,\overset{8}{9}\overset{9}{0}6 \\ -1,088 \\ \hline 1,818 \end{array}$ $\begin{array}{r} \overset{1}{1},\overset{1}{0}88 \\ +1,818 \\ \hline 2,906 \end{array}$

6. $\begin{array}{r} 6,\overset{5}{4}\overset{3}{2}9 \\ -3,587 \\ \hline 2,842 \end{array}$ $\begin{array}{r} \overset{1}{3},\overset{1}{5}87 \\ +2,842 \\ \hline 6,429 \end{array}$

7. $\begin{array}{r} 3,\overset{3}{4}\overset{9}{0}5 \\ - \ \ 159 \\ \hline 3,246 \end{array}$ $\begin{array}{r} \overset{1}{1}\overset{1}{5}9 \\ +3,246 \\ \hline 3,405 \end{array}$

8. $\begin{array}{r} \overset{6}{7},\overset{9}{0}01 \\ -5,991 \\ \hline 1,010 \end{array}$ $\begin{array}{r} \overset{1}{5},\overset{1}{9}91 \\ +1,010 \\ \hline 7,001 \end{array}$

9. $\begin{array}{r} \overset{8}{9},\overset{2}{3}\overset{4}{2}4 \\ -8,358 \\ \hline 966 \end{array}$ $\begin{array}{r} \overset{1}{8},\overset{1}{3}\overset{1}{5}8 \\ + \ \ 966 \\ \hline 9,324 \end{array}$

10. $2,004 - 1,492 = 512$ years
11. $\$3,600 - \$150 = \$3,450$
12. $2,562 - 1,600 = 962$ bushels

Systematic Review 26D

1. $4,305 \quad 289$
    $-\ 289 \quad +4,016$
    $\overline{4,016} \quad \overline{4,305}$

2. $5,840 \quad 3,914$
    $-3,914 \quad +1,296$
    $\overline{1,926} \quad \overline{5,840}$

3. $7,013 \quad 1,523$
    $-1,523 \quad +5,490$
    $\overline{5,490} \quad \overline{7,013}$

4. $200 \quad 34$
    $-\ 34 \quad +166$
    $\overline{166} \quad \overline{200}$

5. $772 \quad 416$
    $-416 \quad +356$
    $\overline{356} \quad \overline{772}$

6. $29 \quad 15$
    $-15 \quad +14$
    $\overline{14} \quad \overline{29}$

7. 卌 ||||

8. 卌 卌 卌 卌 卌 卌 |

9. 卌 卌 ||||

10. 6:50

11. 3:14

12. Tuesday

13. twelfth

14. $12+12+12+12+12 = 60"$

15. $451+385 = 836$ traveled
    $1,000 - 836 = 164$ to go

Systematic Review 26E

1. $1,800 \quad 176$
    $-\ 176 \quad +1,624$
    $\overline{1,624} \quad \overline{1,800}$

2. $6,950 \quad 4,867$
    $-4,867 \quad +2,083$
    $\overline{2,083} \quad \overline{6,950}$

3. $2,093 \quad 2,075$
    $-2,075 \quad +\ 18$
    $\overline{18} \quad \overline{2,093}$

4. $741 \quad 53$
    $-\ 53 \quad +688$
    $\overline{688} \quad \overline{741}$

5. $500 \quad 211$
    $-211 \quad +289$
    $\overline{289} \quad \overline{500}$

6. $48 \quad 19$
    $-19 \quad +29$
    $\overline{29} \quad \overline{48}$

7. 卌

8. 卌 卌 卌 卌 ||||

9. 卌 卌 ||

10. 11:39

11. 9:26

12. October

13. seventh

14. $673+415 = 1,088$
    $1,088 - 223 = 865$ acorns

15. $10+10+10+10 = 40"$

Systematic Review 26F

1. $8,901 \quad 792$
    $-\ 792 \quad +8,109$
    $\overline{8,109} \quad \overline{8,901}$

2. $9,012 \quad 8,999$
    $-8,999 \quad +\ 13$
    $\overline{13} \quad \overline{9,012}$

3.
$$\begin{array}{r} {}^{1}\cancel{2}\,{}^{\cancel{2}}\cancel{3}\,{}^{14} \\ 1,234 \\ -1,045 \\ \hline 189 \end{array} \qquad \begin{array}{r} {}^{11} \\ 1,045 \\ +\;\;189 \\ \hline 1,234 \end{array}$$

4.
$$\begin{array}{r} {}^{6} \\ 6\cancel{7}\,\cancel{0} \\ -\;\;69 \\ \hline 601 \end{array} \qquad \begin{array}{r} {}^{1} \\ 69 \\ +601 \\ \hline 670 \end{array}$$

5.
$$\begin{array}{r} {}^{6}\;{}^{5} \\ \cancel{7}\,\cancel{6}\,\cancel{5} \\ -578 \\ \hline 187 \end{array} \qquad \begin{array}{r} {}^{11} \\ 578 \\ +187 \\ \hline 765 \end{array}$$

6.
$$\begin{array}{r} {}^{2} \\ \cancel{3}\,\cancel{2} \\ -25 \\ \hline 7 \end{array} \qquad \begin{array}{r} {}^{1} \\ 25 \\ +\;\;7 \\ \hline 32 \end{array}$$

7. 𝍷𝍷𝍷𝍷𝍷 𝍷𝍷𝍷𝍷𝍷 𝍷

8. 𝍷𝍷𝍷𝍷𝍷 𝍷𝍷𝍷𝍷𝍷 𝍷𝍷𝍷𝍷𝍷 𝍷𝍷𝍷𝍷𝍷 𝍷𝍷𝍷𝍷𝍷 𝍷𝍷𝍷𝍷𝍷 𝍷𝍷𝍷

9. 𝍷𝍷𝍷𝍷𝍷 𝍷𝍷𝍷𝍷𝍷 𝍷𝍷𝍷𝍷𝍷 𝍷𝍷𝍷𝍷𝍷 𝍷𝍷𝍷𝍷𝍷 𝍷

10. 11 : 31

11. 7 : 41

12. Thursday

13. sixth

14. $2 + 4 + 5 + 6 + 8 = 25$ calls

15. $6 + 3 + 6 + 3 = 18"$

Lesson Practice 27A

1. done

2.
$$\begin{array}{r} {}^{8}\;{}^{9} \\ \$\cancel{9}.\cancel{0}\,\cancel{0} \\ -\;2.67 \\ \hline \$6.33 \end{array}$$

3.
$$\begin{array}{r} \$0.46 \\ -0.10 \\ \hline \$0.36 \end{array}$$

4.
$$\begin{array}{r} {}^{5} \\ \$6.\cancel{1}4 \\ -\;1.21 \\ \hline \$4.93 \end{array}$$

5.
$$\begin{array}{r} {}^{3} \\ \$2.\cancel{4}5 \\ -\;1.38 \\ \hline \$1.07 \end{array}$$

6.
$$\begin{array}{r} {}^{7} \\ \$0.8\,\cancel{0} \\ -0.24 \\ \hline \$0.56 \end{array}$$

7.
$$\begin{array}{r} {}^{4} \\ \$5.\cancel{1}9 \\ -\;3.72 \\ \hline \$1.47 \end{array}$$

8.
$$\begin{array}{r} {}^{5}\;{}^{9} \\ \$2\cancel{6}.\cancel{0}\,\cancel{0} \\ -11.92 \\ \hline \$14.08 \end{array}$$

9.
$$\begin{array}{r} {}^{3} \\ \$34.\cancel{4}\,\cancel{1}7 \\ -22.09 \\ \hline \$12.38 \end{array}$$

10. $\$7.15 - \$2.98 = \$4.17$

11. $\$9.46 - \$4.91 = \$4.55$

12. $\$45.50 - \$34.99 = \$10.51$

Lesson Practice 27B

1.
$$\begin{array}{r} {}^{5} \\ \$2.\cancel{6}\,\cancel{1}5 \\ -0.38 \\ \hline \$2.27 \end{array}$$

2.
$$\begin{array}{r} {}^{6}\;{}^{4} \\ \$\cancel{7}.\cancel{2}\,\cancel{5} \\ -\;1.89 \\ \hline \$5.36 \end{array}$$

3.
$$\begin{array}{r} \$0.10 \\ -0.08 \\ \hline \$0.02 \end{array}$$

4.
$$\begin{array}{r} {}^{9} \\ \$1.\cancel{0}\,\cancel{0} \\ -0.77 \\ \hline \$0.23 \end{array}$$

5.
$$\begin{array}{r} {}^{5} \\ \$6.\cancel{1}9 \\ -\;3.58 \\ \hline \$2.61 \end{array}$$

6.
$$\begin{array}{r} {}^{8} \\ \$0.\cancel{9}\,\cancel{3} \\ -0.45 \\ \hline \$0.48 \end{array}$$

7. $$\begin{array}{r} \overset{3}{\$2.4\,7} \\ -1.0\,9 \\ \hline \$1.3\,8 \end{array}$$

8. $$\begin{array}{r} \overset{4}{\$3\,5.\,6\,0} \\ -2\,1.\,9\,0 \\ \hline \$1\,3.\,7\,0 \end{array}$$

9. $$\begin{array}{r} \overset{6}{\$\,7}\,{}^{1}4.\overset{7}{8}\,{}^{1}2 \\ -\,3\,6.\,2\,5 \\ \hline \$\,3\,8.\,5\,7 \end{array}$$

10. $31.16 - $25.69 = $5.47

11. $39.67 - $20.75 = $18.92

12. $13.05 - $0.36 = $12.69

Lesson Practice 27C

1. $$\begin{array}{r} \overset{4}{\$5}.\overset{0}{1}\,3 \\ -0.\,4\,8 \\ \hline \$4.\,6\,5 \end{array}$$

2. $$\begin{array}{r} \$8.\overset{0}{1}\,5 \\ -7.\,0\,9 \\ \hline \$1.\,0\,6 \end{array}$$

3. $$\begin{array}{r} \$0.\overset{6}{7}\,3 \\ -0.\,1\,7 \\ \hline \$0.\,5\,6 \end{array}$$

4. $$\begin{array}{r} \overset{2}{\$3}.\overset{9}{0}\,0 \\ -0.\,8\,1 \\ \hline \$2.\,1\,9 \end{array}$$

5. $$\begin{array}{r} \$9.\overset{2}{3}\,2 \\ -6.\,1\,4 \\ \hline \$3.\,1\,8 \end{array}$$

6. $$\begin{array}{r} \$0.8\,6 \\ -0.5\,5 \\ \hline \$0.3\,1 \end{array}$$

7. $$\begin{array}{r} \$4.\overset{6}{7}\,5 \\ -2.0\,6 \\ \hline \$2.6\,9 \end{array}$$

8. $$\begin{array}{r} \overset{9}{\$2}\overset{1\text{ to }11}{0.1}\,5 \\ -\,1\,3.\,2\,1 \\ \hline \$6.\,9\,4 \end{array}$$

9. $$\begin{array}{r} \overset{7\;10}{\$9\,8.1}\,7 \\ -\,2\,5.\,1\,8 \\ \hline \$7\,2.\,9\,9 \end{array}$$

10. $10.00 - $1.25 = $8.75

11. $53.95 - $45.00 = $8.95

12. $0.61 - $0.45 = $0.16

Systematic Review 27D

1. $$\begin{array}{r} \overset{8}{\$8.9}\,7 \\ -0.\,7\,6 \\ \hline \$8.\,1\,5 \end{array}$$

2. $$\begin{array}{r} \$6.8\,2 \\ -1.6\,1 \\ \hline \$5.2\,1 \end{array}$$

3. $$\begin{array}{r} \overset{6\ \ 6}{7,7}\,12 \\ -5,\,8\,7\,2 \\ \hline 1,\,8\,4\,0 \end{array}$$

4. $$\begin{array}{r} \overset{1\,1}{1\,8\,4} \\ +\ \ 6\,8 \\ \hline 2\,5\,2 \end{array}$$

5. $$\begin{array}{r} \overset{1\,1}{2\,5\,5} \\ +1\,7\,7 \\ \hline 4\,3\,2 \end{array}$$

6. $$\begin{array}{r} \overset{1}{3,0\,1\,1} \\ +1,\,8\,9\,5 \\ \hline 4,\,9\,0\,6 \end{array}$$

7. first — Wednesday
8. second — Friday
9. third — Saturday
10. fourth — Sunday
11. fifth — Thursday
12. sixth — Monday
13. seventh — Tuesday

14. 2, 4, 6, 8, 10, 12, 14, 16, 18, 20
15. 17 birds
16. $5 + $5 + $5 + $5 = $20
   $20.00 - $6.95 = $13.05

Systematic Review 27E

1. $4.4\overset{3}{\cancel{2}}$
   -0.39
   $\overline{\$4.03}$

2. $1.\overset{1}{\cancel{2}}\cancel{0}$
   -1.18
   $\overline{\$0.02}$

3. $\overset{3}{\cancel{4}},503$
   $-2,901$
   $\overline{1,602}$

4. $\overset{1}{8}36$
   $+\ 17$
   $\overline{853}$

5. $\$\ 3.45$
   $+\ 7.18$
   $\overline{\$10.63}$

6. $\overset{1}{1},\overset{1}{8}19$
   $+4,428$
   $\overline{6,247}$

7. first — June
8. second — April
9. third — January
10. fourth — March
11. fifth — February
12. sixth — May
13. 5, 10, 15, 20, 25, 30, 35, 40, 45, 50
14. 30 days
15. Drew: $5.00 + 6.50 = $11.50
   $11.50 > $10.20; Drew earned more.

Systematic Review 27F

1. $\$3.\overset{0}{\cancel{1}}{}^{1}4$
   -1.09
   $\overline{\$2.05}$

2. $\$2.\overset{5}{\cancel{6}}{}^{1}2$
   -1.38
   $\overline{\$1.24}$

3. $\$1\overset{8}{\cancel{9}}.{}^{1}07$
   $-\ 15.21$
   $\overline{\$\ 3.86}$

4. $\overset{1}{\ }977$
   $+\ 13$
   $\overline{990}$

5. $\$\ 6.54$
   $+\ 9.54$
   $\overline{\$16.08}$

6. $\overset{2\ 1\ 1}{1,764}$
   $1,913$
   $+2,384$
   $\overline{6,061}$

7. seventh — August
8. eighth — October
9. ninth — December
10. tenth — July
11. eleventh — November
12. twelfth — September
13. 10, 20, 30, 40, 50, 60, 70, 80, 90, 100
14. ||||| ||||| ||||| ||||| ||||| ||||| |
15. 35 + 25 + 19 = 79'
   79 - 10 = 69'

Lesson Practice 28A

1. done
2. $\ \ 84,528 \qquad 64,025$
   $-64,025 \qquad +20,503$
   $\overline{\ \ 20,503} \qquad \overline{84,528}$

3.
$$
\begin{array}{r}
\overset{4}{3}\overset{8}{5},\overset{}{7}\overset{9}{9}\overset{}{4} \\
-31,486 \\
\hline 3,708
\end{array}
\qquad
\begin{array}{r}
\overset{1}{3}1,\overset{1}{4}86 \\
+ 3,708 \\
\hline 35,194
\end{array}
$$

4.
$$
\begin{array}{r}
\overset{1}{4}\overset{2}{2},\overset{}{3}\overset{}{5}5 \\
-21,472 \\
\hline 20,883
\end{array}
\qquad
\begin{array}{r}
\overset{1}{2}\overset{1}{1},472 \\
+20,883 \\
\hline 42,355
\end{array}
$$

5.
$$
\begin{array}{r}
\overset{5}{7}1,\overset{\cancel{6}}{6}\overset{}{2}\overset{\cancel{1}}{1} \\
-41,573 \\
\hline 30,048
\end{array}
\qquad
\begin{array}{r}
\overset{1}{4}\overset{1}{1},573 \\
+30,048 \\
\hline 71,621
\end{array}
$$

6.
$$
\begin{array}{r}
\overset{3}{5}6,\overset{9}{4}\overset{}{0}\overset{}{8} \\
-24,379 \\
\hline 32,029
\end{array}
\qquad
\begin{array}{r}
\overset{1}{2}\overset{1}{4},379 \\
+32,029 \\
\hline 56,408
\end{array}
$$

7. 45,900 – 22,175 = 23,725 people
8. 31,231 – 19,452 = 11,779 tadpoles
9. 25,000 – 19,000 = 6,000 miles
10. $45,575 – $38,196 = $7,379

Lesson Practice 28B

1.
$$
\begin{array}{r}
\overset{5}{6}\overset{4}{8},\overset{\cancel{0}}{5}\overset{}{1}\overset{}{1} \\
-19,333 \\
\hline 49,178
\end{array}
\qquad
\begin{array}{r}
\overset{1}{1}9,\overset{1}{3}33 \\
+49,178 \\
\hline 68,511
\end{array}
$$

2.
$$
\begin{array}{r}
25,3\overset{7}{8}2 \\
-11,255 \\
\hline 14,127
\end{array}
\qquad
\begin{array}{r}
1\overset{1}{1},255 \\
+14,127 \\
\hline 25,382
\end{array}
$$

3.
$$
\begin{array}{r}
47,4\overset{5}{6}0 \\
-33,419 \\
\hline 14,041
\end{array}
\qquad
\begin{array}{r}
3\overset{1}{3},419 \\
+14,041 \\
\hline 47,460
\end{array}
$$

4.
$$
\begin{array}{r}
\overset{6}{7}\overset{9}{0},\overset{9}{0}\overset{9}{0}\overset{}{0} \\
-19,999 \\
\hline 50,001
\end{array}
\qquad
\begin{array}{r}
1\overset{1}{9},\overset{1}{9}\overset{1}{9}9 \\
+50,001 \\
\hline 70,000
\end{array}
$$

5.
$$
\begin{array}{r}
65,2\overset{3}{4}2 \\
-21,135 \\
\hline 44,107
\end{array}
\qquad
\begin{array}{r}
2\overset{1}{1},135 \\
+44,107 \\
\hline 65,242
\end{array}
$$

6.
$$
\begin{array}{r}
\overset{5}{5}4,\overset{}{9}\overset{\cancel{6}}{6}\overset{}{7} \\
-42,718 \\
\hline 12,249
\end{array}
\qquad
\begin{array}{r}
4\overset{1}{2},718 \\
+12,249 \\
\hline 54,967
\end{array}
$$

7. 10,752 – 6,834 = 3,918 people
8. 55,780 – 29,592 = 26,188 plants
9. 63,360 – 31,680 = 31,680 inches
10. $22,980 – $16,899 = $6,081

Lesson Practice 28C

1.
$$
\begin{array}{r}
13,\overset{4}{5}\overset{\cancel{1}}{2}\overset{}{2} \\
-12,048 \\
\hline 1,474
\end{array}
\qquad
\begin{array}{r}
1\overset{1}{2},048 \\
+ 1,474 \\
\hline 13,522
\end{array}
$$

2.
$$
\begin{array}{r}
7\overset{8}{9},\overset{}{1}47 \\
-24,312 \\
\hline 54,835
\end{array}
\qquad
\begin{array}{r}
2\overset{1}{4},312 \\
+54,835 \\
\hline 79,147
\end{array}
$$

3.
$$
\begin{array}{r}
53,\overset{5}{6}\overset{3}{4}\overset{}{1} \\
-20,465 \\
\hline 33,176
\end{array}
\qquad
\begin{array}{r}
2\overset{1}{0},465 \\
+33,176 \\
\hline 53,641
\end{array}
$$

4.
$$
\begin{array}{r}
8\overset{7}{4},\overset{3}{6}\overset{6}{7}\overset{}{0} \\
-35,901 \\
\hline 48,769
\end{array}
\qquad
\begin{array}{r}
3\overset{1}{5},\overset{1}{9}\overset{1}{0}1 \\
+48,769 \\
\hline 84,670
\end{array}
$$

5.
$$
\begin{array}{r}
1\overset{\cancel{2}}{3},\overset{4}{5}06 \\
- 9,951 \\
\hline 3,555
\end{array}
\qquad
\begin{array}{r}
\overset{1}{9},\overset{1}{9}\overset{1}{5}1 \\
+ 3,555 \\
\hline 13,506
\end{array}
$$

6.
$$
\begin{array}{r}
4\overset{0}{1},\overset{9}{0}03 \\
-10,523 \\
\hline 30,480
\end{array}
\qquad
\begin{array}{r}
1\overset{1}{0},523 \\
+30,480 \\
\hline 41,003
\end{array}
$$

7. 23,295 – 19,761 = 3,534 miles
8. 53,600 – 48,096 = 5,504 blocks
9. 60,000 – 22,000 = 38,000 fish
10. $95,899 – $26,900 = $68,999

Systematic Review 28D

1.
```
  5 9 ʰ
 96,021      45,635
-45,635     +50,386
 50,386      96,021
```

2.
```
  7 9 9 9
 80,000      79,998
-79,998     +     2
      2      80,000
```

3.
```
$50.75
-10.25
$40.50
```

4.
```
    8 ᵇ
$39.13
-22.46
$16.67
```

5.
```
    8 ᵇ
$19.15
- 7.99
$11.16
```

6. 3 < 4

7. 7 > 6

8. 9 = 9

9.
```
   2 1
  125
  285
   44
+ 161
  615
```

10.
```
  1 1
  315
  200
  394
+ 155
 1,064
```

11.
```
  1 2
  449
  131
  503
+  67
 1,150
```

12. Sunday

13. third

14. $123.45 + $210.13 + $75.21 + $103.82 = $512.61

15. 100 – 99 = 1 sheep

Systematic Review 28E

1.
```
  0 ᵇ
 11,435      10,682
-10,682     +   753
    753      11,435
```

2.
```
  3   6 ᵇ
 40,734      27,156
-27,156     +13,578
 13,578      40,734
```

3.
```
   1 ᵇ ᵇ
$21.11
-19.89
$ 1.22
```

4.
```
    7
$78.04
-35.40
$42.64
```

5.
```
    1 9
$62.00
-51.19
$10.81
```

6. 14 = 14

7. 2 < 3

8. 63 > 36

9.
```
   1
  704
  300
   26
+   9
 1,039
```

10.
```
  1 2
  241
  119
  562
+ 508
 1,430
```

11.
```
  1 1
  325
  763
  125
+  46
 1,259
```

12. August

13. sixth

14. 10,456 – 8,912 = 1,544 spiderlings

15. 25 + 89 + 5 = 119 animals

Systematic Review 28F

1.
$$\begin{array}{r} 79,3\overset{2}{\cancel{3}}0\overset{1}{4} \\ -61,082 \\ \hline 18,222 \end{array}$$
$$\begin{array}{r} 61,082 \\ +18,222 \\ \hline 79,304 \end{array}$$

2.
$$\begin{array}{r} \overset{4}{\cancel{5}}\overset{14}{\cancel{5}},\overset{9}{\cancel{0}}\overset{9}{\cancel{0}}\cancel{0} \\ -48,123 \\ \hline 6,877 \end{array}$$
$$\begin{array}{r} \overset{1}{4}\overset{1}{8},\overset{1}{1}2\overset{1}{3} \\ + 6,877 \\ \hline 55,000 \end{array}$$

3.
$$\begin{array}{r} \$17.99 \\ - 5.15 \\ \hline \$12.84 \end{array}$$

4.
$$\begin{array}{r} \$4\overset{3}{\cancel{3}}.\overset{2}{\cancel{6}}\overset{5}{\cancel{0}} \\ - 13.79 \\ \hline \$29.81 \end{array}$$

5.
$$\begin{array}{r} \$28.\overset{6}{\cancel{7}}\overset{14}{4} \\ - 13.25 \\ \hline \$15.49 \end{array}$$

6. $42 < 78$

7. $16 > 11$

8. $3 = 3$

9.
$$\begin{array}{r} ^{1\,1} \\ 910 \\ 262 \\ 366 \\ +148 \\ \hline 1,686 \end{array}$$

10.
$$\begin{array}{r} ^{2\,1} \\ 355 \\ 176 \\ 755 \\ + 23 \\ \hline 1,309 \end{array}$$

11.
$$\begin{array}{r} ^{1\,1} \\ 431 \\ 529 \\ 611 \\ +576 \\ \hline 2,147 \end{array}$$

12. Wednesday

13. fifth

14. $34.00 + 25.00 = 59.00$
$59.00 - 29.98 = 29.02$

15. $25 + 17 + 30 + 8 = 80$ cars

Lesson Practice 29A

1. done

2. 5 (See Example 1 on student page for filled-in numbers on gauge.)

3. 10

4. 50

5. 10

6. 70

7. done

8.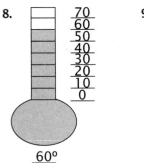
60°

9.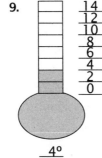
4°

Lesson Practice 29B

1. done

2. 100°

3. 600°

4. 30

5. 90

6. 60

7.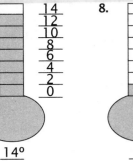
14°

8.
35
30
25
20
15
10
5
0
5°

9.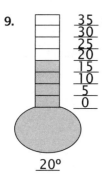

20°

Lesson Practice 29C

1. 20
2. 0
3. 15
4. 20
5. 80
6. 40

7.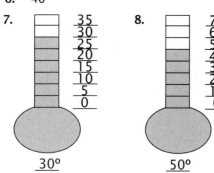

30°   50°

9.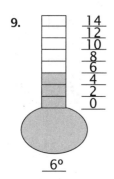

6°

Systematic Review 29D

1. 5
2. 300°

3.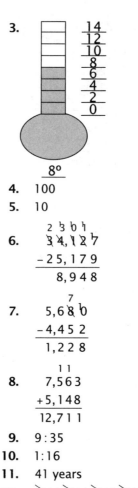

8°

4. 100
5. 10

6.
$$\begin{array}{r} 2\ \overset{3}{\cancel{3}}\ \overset{6}{\cancel{4}}\ \overset{0}{\cancel{1}}\ \overset{7}{\cancel{2}}\ 7 \\ 3\,4,1\,2\,7 \\ -2\,5,1\,7\,9 \\ \hline 8,9\,4\,8 \end{array}$$

7.
$$\begin{array}{r} \overset{7}{5,6\,\overset{7}{8}\,0} \\ -4,4\,5\,2 \\ \hline 1,2\,2\,8 \end{array}$$

8.
$$\begin{array}{r} \overset{1\ 1}{7,5\,6\,3} \\ +5,1\,4\,8 \\ \hline 1\,2,7\,1\,1 \end{array}$$

9. 9:35
10. 1:16
11. 41 years
12. 𝍸𝍸 𝍸𝍸 𝍸𝍸 𝍸𝍸 𝍸𝍸 𝍸𝍸 𝍸𝍸 𝍸𝍸 I

Systematic Review 29E

1. 20
2. 500°

3.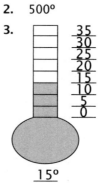

15°

4. 50

5. 0

6.
$$\begin{array}{r} 6{}^{4} \\ 1\cancel{7},\cancel{5}39 \\ -15,659 \\ \hline 1,880 \end{array}$$

7.
$$\begin{array}{r} 69 \\ \cancel{7},\cancel{0}58 \\ -3,172 \\ \hline 3,886 \end{array}$$

8.
$$\begin{array}{r} 11 \\ 4,264 \\ +2,791 \\ \hline 7,055 \end{array}$$

9. left

10. 1976 – 200 = 1776; seventh

8.
$$\begin{array}{r} 111 \\ 6,456 \\ 2,434 \\ +1,849 \\ \hline 10,739 \end{array}$$

9. 11:43

10. 8:26

11. left

12. step 1: 1976 – 200 = 1776
    step 2: Subtract 1776 from current year.
    Answers will vary according to year.
    There is more than one correct way
    to solve this.

Systematic Review 29F

1. 10

2. 200°

3.
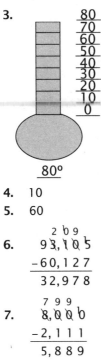

80°

4. 10

5. 60

6.
$$\begin{array}{r} 2\cancel{0}9 \\ 9\cancel{3},\cancel{1}\cancel{0}5 \\ -60,127 \\ \hline 32,978 \end{array}$$

7.
$$\begin{array}{r} 799 \\ \cancel{8},\cancel{0}\cancel{0}\cancel{0} \\ -2,111 \\ \hline 5,889 \end{array}$$

Lesson Practice 30A

1. February

2. December

3. 2"

4. January

5.
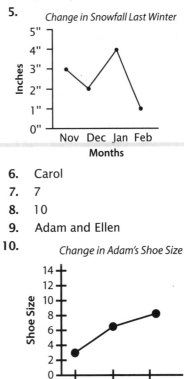

6. Carol

7. 7

8. 10

9. Adam and Ellen

10.

tice 30B

...sday

...ay

...oughnuts

4.   300 Doughnuts

5.

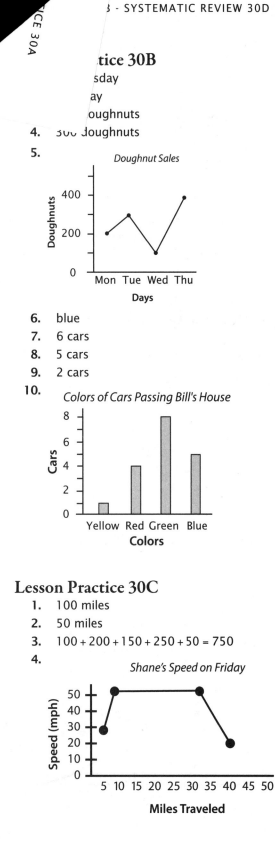

Doughnut Sales

6.   blue
7.   6 cars
8.   5 cars
9.   2 cars
10.

Colors of Cars Passing Bill's House

Lesson Practice 30C

1.   100 miles
2.   50 miles
3. $100 + 200 + 150 + 250 + 50 = 750$
4.

Shane's Speed on Friday

5.   16 books
6.   12 books
7. $8 + 12 + 10 + 16 = 46$ books
8.

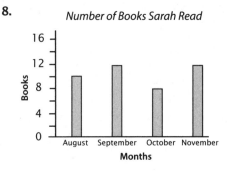

Number of Books Sarah Read

Systematic Review 30D

1.   Thursday and Friday
2.   Wednesday
3. $3 + 3 + 2 + 4 + 4 = 16$ miles
4.

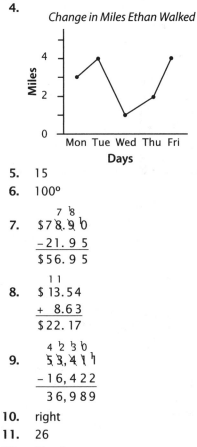

Change in Miles Ethan Walked

5.   15
6. $100°$

7.
$$\begin{array}{r} \overset{7}{}\overset{8}{} \\ \$7\,\overset{}{8}.\overset{}{9}\,0 \\ -\ 2\,1.\,9\,5 \\ \hline \$5\,6.\,9\,5 \end{array}$$

8.
$$\begin{array}{r} \overset{1\ 1}{} \\ \$\ 1\,3.\,5\,4 \\ +\ \ 8.\,6\,3 \\ \hline \$2\,2.\,1\,7 \end{array}$$

9.
$$\begin{array}{r} \overset{4\ 2\ 3\ 0}{5\,3,4\,1\,1} \\ -\ 1\,6,4\,2\,2 \\ \hline 3\,6,9\,8\,9 \end{array}$$

10. right
11. 26
12. March

Systematic Review 30E

1. May
2. July
3. 5"
4.

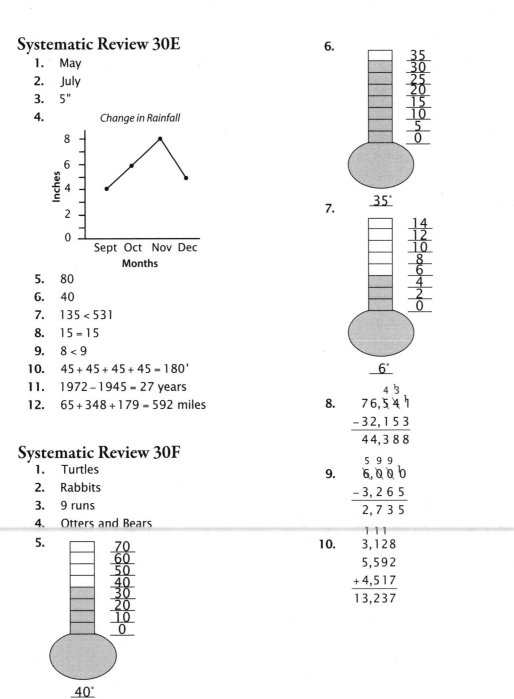

Change in Rainfall

5. 80
6. 40
7. 135 < 531
8. 15 = 15
9. 8 < 9
10. 45 + 45 + 45 + 45 = 180'
11. 1972 – 1945 = 27 years
12. 65 + 348 + 179 = 592 miles

Systematic Review 30F

1. Turtles
2. Rabbits
3. 9 runs
4. Otters and Bears
5.   40°

6.   35°

7.   6°

8.
$$\begin{array}{r} {}^{4}\,{}^{3} \\ 76,5\,4\,\cancel{7} \\ -\,32,1\,5\,3 \\ \hline 44,3\,8\,8 \end{array}$$

9.
$$\begin{array}{r} {}^{5}\,{}^{9}\,{}^{9} \\ \cancel{6},\cancel{0}\,\cancel{0}\,0 \\ -\,3,2\,6\,5 \\ \hline 2,7\,3\,5 \end{array}$$

10.
$$\begin{array}{r} {}^{1}\,{}^{1}\,{}^{1} \\ 3,1\,2\,8 \\ 5,5\,9\,2 \\ +\,4,5\,1\,7 \\ \hline 13,2\,3\,7 \end{array}$$

Appendix A1

Check student drawings for each of the following.

1. triangle
2. pentagon
3. quadrilateral
4. hexagon
5. pentagon

Appendix B1

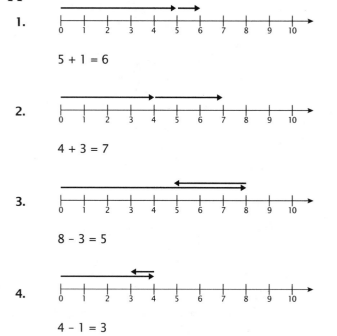

1.

5 + 1 = 6

2.

4 + 3 = 7

3.

8 − 3 = 5

4.

4 − 1 = 3

Test Solutions

Lesson Test 1

1. 264; "two hundred sixty-four"
2. 39; "thirty-nine"
3. 4 hundreds, 2 tens, and 1 unit; "four hundred twenty-one"
4. 2 hundreds and 7 units; "two hundred seven"
5. $0 + 7 = 7$
6. $8 + 3 = 11$
7. $9 + 7 = 16$
8. $6 + 5 = 11$
9. $4 + 4 = 8$
10. $3 + 6 = 9$
11. $8 + 2 = 10$
12. $4 + 7 = 11$
13. $5 + 5 = 10$
14. $3 + 5 = 8$
15. $6 + 4 = 10$
16. $5 + 7 = 12$
17. $9 + 1 = 10$
18. $9 + 6 = 15$

Lesson Test 2

1. 16, 20, 34
2. 19, 49, 99
3. 108, 83, 82
4. 111, 101, 11
5. 22, 23, 24, 25, 26
6. 1 hundred and 4 tens; "one hundred forty"
7. 2 tens and 2 units; "twenty-two"
8. $4 + 0 = 4$
9. $2 + 7 = 9$
10. $5 + 9 = 14$
11. $3 + 6 = 9$
12. $2 + 3 = 5$
13. $7 + 7 = 14$
14. $9 + 1 = 10$
15. 14 is less than 24; Chad
16. $8 + 9 = 17$ arrowheads

Lesson Test 3

1. $11 > 7$
2. $9 = 9$
3. $19 < 91$
4. $111 > 101$
5. $13 = 13$
6. $9 < 10$
7. 29, 28, 27, 26, 25
8. $2 + 8 = 10$
9. $6 + 9 = 15$
10. $3 + 4 = 7$
11. $0 + 2 = 2$
12. $7 + 8 = 15$
13. $4 + 7 = 11$
14. $9 + 4 = 13$
15. $12 < 14$
16. $5 + 0 = 5$ dollars

Lesson Test 4

1. 90
2. 30
3. 50

4.
$$\begin{array}{r} (50) \\ +(20) \\ \hline (70) \end{array}$$

5.
$$\begin{array}{r} (40) \\ +(30) \\ \hline (70) \end{array}$$

6.
$$\begin{array}{r} (30) \\ +(30) \\ \hline (60) \end{array}$$

7.
$$\begin{array}{r} (80) \\ +(10) \\ \hline (90) \end{array}$$

8.

4	4	8
3	2	5
7	6	13

9.

1	6	7
5	3	8
6	9	15

10. $10 > 7$
11. $399 > 329$
12. $2 + 2 = 4$ muffins
13. $(30) + (30) = (60)$ cans
14. $201 > 21$

Lesson Test 5

1. $600 + 20 + 8$
2. $40 + 9$
3.
$$\begin{array}{rr} 16 & 10+6 \\ +32 & +30+2 \\ \hline 48 & 40+8 \end{array}$$
4.
$$\begin{array}{rr} 380 & 300+80+0 \\ +516 & +500+10+6 \\ \hline 896 & 800+90+6 \end{array}$$
5.
$$\begin{array}{rr} 54 & 50+4 \\ +22 & +20+2 \\ \hline 76 & 70+6 \end{array}$$
6.
$$\begin{array}{rr} 433 & 400+30+3 \\ +425 & +400+20+5 \\ \hline 858 & 800+50+8 \end{array}$$
7.
$$\begin{array}{rr} 34 & (30) \\ +61 & +(60) \\ \hline 95 & (90) \end{array}$$
8.
$$\begin{array}{rr} 12 & (10) \\ +16 & +(20) \\ \hline 28 & (30) \end{array}$$
9. $10 = 10$
10. $49 < 94$
11. $7 + \underline{7} = 14$
12. $5 + \underline{3} = 8$
13. $8 + \underline{8} = 16$
14. $45 + 12 = 57$ animals
15. $132 + 160 = 292$ coins

Lesson Test 6

1. 2, 4, 6, 8, 10, 12, 14, 16, 18, 20
2. 2, 4, 6, $\underline{8}$ cones
3.
$$\begin{array}{rr} 55 & 50+5 \\ +\ 2 & +00+2 \\ \hline 57 & 50+7 \end{array}$$

4.
$$\begin{array}{rr} 106 & 100+00+6 \\ +221 & +200+20+1 \\ \hline 327 & 300+20+7 \end{array}$$
5.
$$\begin{array}{rr} 44 & (40) \\ +13 & +(10) \\ \hline 57 & (50) \end{array}$$
6.
$$\begin{array}{rr} 38 & (40) \\ +21 & +(20) \\ \hline 59 & (60) \end{array}$$
7. $2 + \underline{6} = 8$
8. $5 + \underline{8} = 13$
9. $9 + \underline{9} = 18$
10. $5 + 9 = 14$ trucks
11. 2, 4, $\underline{6}$ hands
12. $40 + 25 = 65$ jelly beans

Lesson Test 7

1.
$$\begin{array}{r} 1 \\ 35 \\ +29 \\ \hline 64 \end{array}$$
2.
$$\begin{array}{r} 1 \\ 16 \\ +67 \\ \hline 83 \end{array}$$
3.
$$\begin{array}{r} 1 \\ 29 \\ +44 \\ \hline 73 \end{array}$$
4.
$$\begin{array}{r} 1 \\ 19 \\ +46 \\ \hline 65 \end{array}$$
5.
$$\begin{array}{r} 1 \\ 54 \\ +\ 7 \\ \hline 61 \end{array}$$
6.
$$\begin{array}{r} 75 \\ +22 \\ \hline 97 \end{array}$$
7. 2, 4, 6, 8, 10, 12, 14, 16, 18, 20
8. $15 > 13$
9. $21 > 12$
10. $801 < 810$

11. $4 + \underline{1} = 5$
12. $9 + \underline{3} = 12$
13. $4 + \underline{3} = 7$
14. $55 + 39 = 94$ quarts
15. $212 + 344 = 556$ animals

Unit Test I

1. 134
   "one hundred thirty-four"
2. 17, 71, 111
3. $11 > 10$
4. $14 = 14$
5. $342 > 324$
6. 20
7. 40
8. 80
9. $5 + \underline{2} = 7$
10. $8 + \underline{8} = 16$
11. $5 + \underline{4} = 9$
12. 2, 4, 6, 8, 10, 12, 14, 16, 18, 20
13.
$$\begin{array}{r} 16 \\ +21 \\ \hline 37 \end{array}$$
14.
$$\begin{array}{r} {\scriptstyle 1} \\ 55 \\ +37 \\ \hline 92 \end{array}$$
15.
$$\begin{array}{r} {\scriptstyle 1} \\ 15 \\ +25 \\ \hline 40 \end{array}$$
16.
$$\begin{array}{r} {\scriptstyle 1} \\ 78 \\ +\ 8 \\ \hline 86 \end{array}$$
17.
$$\begin{array}{r} 234 \\ +142 \\ \hline 376 \end{array}$$
18.
$$\begin{array}{r} 361 \\ +205 \\ \hline 566 \end{array}$$
19. 2, 4, 6, $\underline{8}$ letters
20. $(30) + (40) = (70)$
    $28 + 43 = 71$ cars

21. $415 + 221 = 636$ fish
22. $39 + 15 = 54$ dollars

Lesson Test 8

1. 10, 20, 30, 40, 50, 60, 70, 80, 90, 100
2. 2, 4, 6, 8, 10, 12, 14, 16, 18, 20
3.
$$\begin{array}{r} {\scriptstyle 1} \\ 45 \\ +27 \\ \hline 72 \end{array}$$
4.
$$\begin{array}{r} {\scriptstyle 1} \\ 65 \\ +16 \\ \hline 81 \end{array}$$
5.
$$\begin{array}{r} {\scriptstyle 1} \\ 37 \\ +34 \\ \hline 71 \end{array}$$
6.
$$\begin{array}{r} 109 \\ +290 \\ \hline 399 \end{array}$$
7.
$$\begin{array}{r} 477 \\ +122 \\ \hline 599 \end{array}$$
8.
$$\begin{array}{r} 834 \\ +\ 45 \\ \hline 879 \end{array}$$
9. $4 + 4 = 8$
10. $5 + \underline{9} = 14$
11. $7 + \underline{3} = 10$
12. 10, 20, 30, 40, 50, 60, $\underline{70}$ toes
13. $4 + 1 = 5$ dimes
    10, 20, 30, 40, $\underline{50}$¢
14. $6 + 1 = 7$
    $7 + 3 = 10$ stories
15. $34 + 47 = 81$ children

Lesson Test 9

1. 5, 10, 15, 20, 25, 30, 35, 40, 45, 50
2. 10, 20, 30, 40, 50, 60, 70, 80, 90, 100
3.
$$\begin{array}{r} {\scriptstyle 1} \\ 26 \\ +55 \\ \hline 81 \end{array}$$

4.
$$\begin{array}{r} \overset{1}{6\,8} \\ +8 \\ \hline 7\,6 \end{array}$$

5.
$$\begin{array}{r} \overset{1}{1\,3} \\ +4\,9 \\ \hline 6\,2 \end{array}$$

6.
$$\begin{array}{r} 1\,1\,6 \\ +4\,3\,1 \\ \hline 5\,4\,7 \end{array}$$

7.
$$\begin{array}{r} 6\,9\,1 \\ +3\,0\,5 \\ \hline 9\,9\,6 \end{array}$$

8.
$$\begin{array}{r} 5\,0\,0 \\ +2\,1\,6 \\ \hline 7\,1\,6 \end{array}$$

9. $7 + \underline{6} = 13$
10. $5 + \underline{4} = 9$
11. $5 + \underline{0} = 5$
12. penny
13. nickel
14. dime
15. 5, 10, 15, 20, 25, <u>30</u> rocks

Lesson Test 10

1. $2.34
   "two dollars and thirty-four cents"
2. 1 dollar, 2 dimes, and 8 pennies
   "one dollar and twenty-eight cents"
3. 2 dollars and 6 pennies
   "two dollars and six cents"
4. 3 dollars, 4 dimes, and one penny
   "three dollars and forty-one cents"
5. 4 dollars, 1 dime, and 5 pennies
   "four dollars and fifteen cents"
6. 5, 10, 15, 20, 25¢

7.
$$\begin{array}{r} \overset{1}{6\,3} \\ +1\,7 \\ \hline 8\,0 \end{array}$$

8.
$$\begin{array}{r} 2\,3\,4 \\ +1\,4\,5 \\ \hline 3\,7\,9 \end{array}$$

9.
$$\begin{array}{r} \overset{1}{4\,8} \\ +2\,6 \\ \hline 7\,4 \end{array}$$

10. 50
11. 20
12. 30
13. $9.54
14. $3 + 2 = 5$
    $5 + 5 = 10$ pounds
15. 2 dimes = 20¢
    5 nickels = 25¢
    20¢ + 25¢ = 45¢ or $0.45

Lesson Test 11

1. 100
2. 300
3. 500

4.
$$\begin{array}{r} \overset{1}{1\,3\,2} \\ +4\,1\,8 \\ \hline 5\,5\,0 \end{array}$$

5.
$$\begin{array}{r} \overset{1\,1}{6\,7\,9} \\ +2\,7\,6 \\ \hline 9\,5\,5 \end{array}$$

6.
$$\begin{array}{r} \overset{1}{5\,2\,0} \\ +1\,8\,8 \\ \hline 7\,0\,8 \end{array}$$

7.
$$\begin{array}{r} \overset{1}{8\,7} \\ +2\,8 \\ \hline 1\,1\,5 \end{array}$$

8.
$$\begin{array}{r} 4\,5 \\ +4\,2 \\ \hline 8\,7 \end{array}$$

9.
$$\begin{array}{r} \overset{1}{3\,9} \\ +1\,5 \\ \hline 5\,4 \end{array}$$

10. $7 - 2 = 5$
11. $5 - 1 = 4$
12. $8 - 0 = 8$
13. $11 - 9 = 2$

14. $5 - 3 = 2$
15. $10 - 2 = 8$
16. $6 - 1 = 5$
17. $7 - 5 = 2$
18. 5, 10, 15, 20, 25, 30, 35, 40, 45, 50
19. $(500) + (100) = (600)$
    $476 + 125 = 601$ stamps
20. $10 - 6 = 4$ years

Lesson Test 12

1.
$$
\begin{array}{r}
\overset{1\ 1}{\ } \\
\$2.46 \\
+1.79 \\
\hline
\$4.25
\end{array}
$$

2.
$$
\begin{array}{r}
\overset{1}{\ } \\
\$3.9\,1 \\
+3.25 \\
\hline
\$7.16
\end{array}
$$

3.
$$
\begin{array}{r}
\overset{1}{\ } \\
\$1.57 \\
+0.38 \\
\hline
\$1.95
\end{array}
$$

4.
$$
\begin{array}{r}
\overset{1}{\ } \\
401 \\
+579 \\
\hline
980
\end{array}
$$

5.
$$
\begin{array}{r}
65 \\
+11 \\
\hline
76
\end{array}
$$

6.
$$
\begin{array}{r}
\overset{1}{\ } \\
22 \\
+\ 8 \\
\hline
30
\end{array}
$$

7. $18 - 9 = 9$
8. $8 - 4 = 4$
9. $13 - 8 = 5$
10. $16 - 9 = 7$
11. $10 - 5 = 5$
12. $12 - 8 = 4$
13. $14 - 7 = 7$
14. $13 - 9 = 4$
15. $1.43; "one dollar and forty-three cents"
16. $1.89 + $2.36 = $4.25

17. 4 dimes = $0.40
    5 nickels = $0.25
    $0.40 + $0.25 = $0.65
18. $15 - 9 = 6$ people

Lesson Test 13

1. $6 + 4 + 7 = 17$
2. $3 + 2 + 7 + 8 = 20$
3.
$$
\begin{array}{r}
\overset{1}{\ } \\
11 \\
42 \\
39 \\
+36 \\
\hline
128
\end{array}
$$

4.
$$
\begin{array}{r}
\overset{1}{\ } \\
\$1.26 \\
+8.47 \\
\hline
\$9.73
\end{array}
$$

5.
$$
\begin{array}{r}
\overset{1}{\ } \\
64\,1 \\
+\ 90 \\
\hline
73\,1
\end{array}
$$

6.
$$
\begin{array}{r}
\overset{1}{\ } \\
58 \\
+\ 2 \\
\hline
60
\end{array}
$$

7. $10 - 7 = 3$
8. $9 - 3 = 6$
9. $10 - 6 = 4$
10. $9 - 4 = 5$
11. $10 - 5 = 5$
12. $9 - 7 = 2$
13. $10 - 2 = 8$
14. $10 - 3 = 7$
15. 2, 4, 6, 8, 10, 12, 14, 16, 18, 20
16. triangle
17. square (a type of rectangle)
18. rectangle
19. $3 + 4 + 6 + 1 = 14$ hours
20. 6 nickels = 30¢
    3 dimes = 30¢
    30¢ + 30¢ + 15¢ = 75¢

Lesson Test 14

1. 3"

2. 6"

3.
   $$\begin{array}{r} {\scriptstyle 1} \\ \$1.34 \\ +2.07 \\ \hline \$3.41 \end{array}$$

4.
   $$\begin{array}{r} {\scriptstyle 1\,1} \\ 866 \\ +\ \ 36 \\ \hline 902 \end{array}$$

5.
   $$\begin{array}{r} {\scriptstyle 1} \\ 26 \\ +44 \\ \hline 70 \end{array}$$

6. $1+1+3+9+7 = 21$

7. $13+4+11+6 = 34$

8. $9 > 8$

9. $13 - 7 = 6$

10. $15 - 9 = 6$

11. $7 - 3 = 4$

12. $14 - 6 = 8$

13. $7 - 4 = 3$

14. $12 - 7 = 5$

15. 3 sides

16. 4 sides

17. $11 - 4 = 7$

    $7 - 2 = 5$ left

18. $12+12+12+12 = 48"$

Lesson Test 15

1. triangle

   $2+4+5 = 11"$

2. square (a type of rectangle)

   $14+14+14+14 = 56"$

3.
   $$\begin{array}{r} {\scriptstyle 1} \\ \$2.22 \\ +3.68 \\ \hline \$5.90 \end{array}$$

4.
   $$\begin{array}{r} {\scriptstyle 1} \\ 771 \\ +135 \\ \hline 906 \end{array}$$

5.
   $$\begin{array}{r} {\scriptstyle 1} \\ 46 \\ +55 \\ \hline 101 \end{array}$$

6. $13 - 9 = 4$

7. $11 - 4 = 7$

8. $13 - 4 = 9$

9. $12 - 5 = 7$

10. $9 - 6 = 3$

11. $11 - 3 = 8$

12. $5+6+5+6 = 22'$

13. $12+12+12 = 36"$

14. 3 dimes = 30¢

    3 nickels = 15¢

    30¢ + 15¢ = 45¢

    45¢ − 5¢ = 40¢ or $0.40

15. $9 - 3 = 6$ pies

Unit Test II

1. $6 = 6$

2. $5 < 7$

3. $117 < 171$

4. 50

5. 90

6. 70

7. 100

8. 400

9. 700

10. 2, 4, 6, 8, 10, 12, 14, 16, 18, 20

11. 5, 10, 15, 20, 25, 30, 35, 40, 45, 50

12. 10, 20, 30, 40, 50, 60, 70, 80, 90, 100

13.
    $$\begin{array}{r} 13 \\ +45 \\ \hline 58 \end{array}$$

14.
    $$\begin{array}{r} {\scriptstyle 1} \\ 64 \\ +28 \\ \hline 92 \end{array}$$

15.
    $$\begin{array}{r} {\scriptstyle 1} \\ 76 \\ +19 \\ \hline 95 \end{array}$$

16.
$$\begin{array}{r} \overset{1\ 1}{}\$3.6\,1 \\ +1.79 \\ \hline \$5.40 \end{array}$$

17.
$$\begin{array}{r} \overset{1}{}4\,5\,2 \\ +256 \\ \hline 708 \end{array}$$

18.
$$\begin{array}{r} \overset{1\ 1}{}9\,6\,8 \\ +\ \ 75 \\ \hline 1{,}043 \end{array}$$

19. $7 + 3 + 2 = 12$

20. $5 + 4 + 5 + 1 = 15$

21.
$$\begin{array}{r} \overset{1}{}6\,5 \\ 12 \\ 18 \\ +\ \ 3 \\ \hline 98 \end{array}$$

22. rectangle
$19 + 21 + 19 + 21 = 80"$

23. $14 - 5 = 9$

24. $18 - 9 = 9$

25. $10 - 4 = 6$

26. $17 - 8 = 9$

27. $7 - 5 = 2$

28. $11 - 6 = 5$

29. $12 + 12 + 12 + 12 = 48"$

30. 3 dimes = 30¢
6 nickels = 30¢
30¢ + 30¢ + 7¢ = 67¢ or $0.67

Lesson Test 16

1. 8,672
"eight thousand,
six hundred seventy-two"

2. 93,145
"ninety-three thousand,
one hundred forty-five"

3. 236,179

4. 11,416

5.
$$\begin{array}{r} \overset{1}{}7\,6\,3 \\ +518 \\ \hline 1{,}281 \end{array}$$

6.
$$\begin{array}{r} \overset{1\ 1}{}3\,5\,4 \\ +956 \\ \hline 1{,}310 \end{array}$$

7.
$$\begin{array}{r} \overset{1\ 1}{}\$3.56 \\ +2.49 \\ \hline \$6.05 \end{array}$$

8.
$$\begin{array}{r} \overset{1}{}5\,6 \\ 44 \\ +98 \\ \hline 198 \end{array}$$

9. $14 - 9 = 5$

10. $13 - 5 = 8$

11. $16 - 7 = 9$

12. $11 - 4 = 7$

13. $10 - 6 = 4$

14. $12 - 8 = 4$

15. 4 nickels = 20¢
3 dimes = 30¢
30¢ > 20¢

16. $4 + 7 = 11$
$11 - 8 = 3$

Lesson Test 17

1. 2,000

2. 10,000

3.
$$\begin{array}{r} \overset{1\ \ \ 1}{}6{,}309 \\ +1{,}712 \\ \hline 8{,}021 \end{array}$$

4.
$$\begin{array}{r} \overset{1}{}8{,}416 \\ +3{,}554 \\ \hline 11{,}970 \end{array}$$

5.
$$\begin{array}{r} \overset{1}{}\$\ 8.92 \\ +\ \ 4.25 \\ \hline \$13.17 \end{array}$$

6. 786,410;
   seven hundred eighty-six thousand,
   four hundred ten
7. $12 + 16 + 20 = 48'$
8. $10 - 3 = 7$
9. $9 - 5 = 4$
10. $15 - 6 = 9$
11. $12 - 4 = 8$
12. $14 - 7 = 7$
13. $7 - 4 = 3$
14. $14 - 5 = 9$ birds
15. $2,176 + 3,402 = 5,578$ ants
16. $\$29 + \$21 + \$9 = \$59$

10. $17 - 8 = 9$
11. $18 - 9 = 9$
12. $7 - 6 = 1$
13. $11 - 8 = 3$
14. 2, 4, 6, 8, 10, 12, 14, 16, 18, 20
15. $5 + 7 + 4 + 10 + 13 = 39$ miles
16. $15 - 7 = 8$ people

Lesson Test 18

1.
$$
\begin{array}{r}
1\,2\\
123\\
678\\
207\\
133\\
+312\\
\hline
1,453
\end{array}
$$

2.
$$
\begin{array}{r}
1\,1\\
662\\
108\\
500\\
543\\
+161\\
\hline
1,974
\end{array}
$$

3.
$$
\begin{array}{r}
2\,2\\
782\\
153\\
269\\
341\\
+807\\
\hline
2,352
\end{array}
$$

4. 871,465
   "eight hundred seventy-one thousand,
   four hundred sixty-five"
5. 56,217
6. $15 - 9 = 6$
7. $10 - 8 = 2$
8. $13 - 9 = 4$
9. $5 - 3 = 2$

Lesson Test 19

1.
$$
\begin{array}{r}
2\,1\,1\\
2,834\\
1,548\\
+3,672\\
\hline
8,054
\end{array}
$$

2.
$$
\begin{array}{r}
1\,1\,1\\
4,506\\
3,294\\
+2,753\\
\hline
10,553
\end{array}
$$

3.
$$
\begin{array}{r}
1\,1\,1\\
6,729\\
5,326\\
+8,361\\
\hline
20,416
\end{array}
$$

4. $10 - 6 = 4$
5. $16 - 9 = 7$
6. $6 - 3 = 3$
7. $11 - 4 = 7$
8. $9 - 5 = 4$
9. $14 - 7 = 7$
10. $11 - 6 = 5$
11. $10 - 5 = 5$
12. $1 = 1$
13. $18 > 8$
14. $209 < 902$
15. $1,367 + 2,079 + 1,534 = 4,980$ trees
16. $10 + 10 + 10 = 30"$
17. 10, 20, 30, 40, 50, 60, <u>70</u> toes
18. $\$0.46 + \$0.05 = \$0.51$

Lesson Test 20

1.
```
  50    20
- 20  + 30
  30    50
```

2.
```
  42    30
- 30  + 12
  12    42
```

3.
```
  85    23
- 23  + 62
  62    85
```

4.
```
  263    112
- 112  + 151
  151    263
```

5.
```
  888    546
- 546  + 342
  342    888
```

6.
```
  374    62
-  62  + 312
  312    374
```

7.
```
   3274
   1690
+ 7216
  12,180
```

8.
```
  5209
  6128
+ 1342
  12,679
```

9.
```
   123
   456
   789
+ 111
  1,479
```

10. 54,971; "fifty-four thousand, nine hundred seventy-one"

11. 2, 4, 6, 8, 10, 12 14, 16, 18, 20

12. 5, 10, 15, 20, 25 30, 35, 40, 45, 50

13. 10, 20, 30, 40, 50 60, 70, 80, 90, 100

14. 29 − 26 = 3 years

15. 48 − 34 = 14 cookies

Lesson Test 21

1. :15

2. :47

3. :36

4. :05

5.
```
  78
-  6
  72
```

6.
```
  562
-  41
  521
```

7.
```
  826
- 603
  223
```

8.
```
$  7.20
+  4.56
$11.76
```

9.
```
    1 1 1
    6,034
+  2,987
    9,021
```

10.
```
    2 1
    594
    346
+  273
   1,213
```

11. 25 − 13 = 12 years old

12. 38 + 107 + 73 = 218 miles

Lesson Test 22

1.
```
  5 1        1
  6̶2    28
- 28  + 34
  34    62
```

2.
```
  2 1        1
  3̶4    15
- 15  + 19
  19    34
```

3.
```
  8 1        1
  9̶3    56
- 56  + 37
  37    93
```

4.
$$
\begin{array}{cc}
\overset{1\ 1}{\cancel{2}} 2 & \overset{1}{1} 8 \\
-1\ 8 & +\ \ 4 \\
\hline
4 & 2\ 2
\end{array}
$$

5.
$$
\begin{array}{cc}
8\ 1\ 7 & 2\ 0\ 5 \\
-2\ 0\ 5 & +6\ 1\ 2 \\
\hline
6\ 1\ 2 & 8\ 1\ 7
\end{array}
$$

6.
$$
\begin{array}{cc}
6\ 2\ 3 & 5\ 1\ 2 \\
-5\ 1\ 2 & +1\ 1\ 1 \\
\hline
1\ 1\ 1 & 6\ 2\ 3
\end{array}
$$

7. $7.33

8. 11,385

9. 1,396

10. :40

11. :59

12. 35 − 9 = 26;
26 − 9 = 17 notes
OR
9 + 9 = 18;
35 − 18 = 17 notes

Unit Test III

1.
$$
\begin{array}{cc}
\overset{5}{\cancel{6}} \overset{1}{8} & \overset{1}{1} 9 \\
-1\ 9 & +4\ 9 \\
\hline
4\ 9 & 6\ 8
\end{array}
$$

2.
$$
\begin{array}{cc}
\overset{3}{\cancel{4}} \overset{1}{4} & 3\ 6 \\
-3\ 6 & +\ \ 8 \\
\hline
8 & 4\ 4
\end{array}
$$

3.
$$
\begin{array}{cc}
\overset{7}{\cancel{8}} \overset{1}{3} & 5\ 8 \\
-5\ 8 & +2\ 5 \\
\hline
2\ 5 & 8\ 3
\end{array}
$$

4.
$$
\begin{array}{cc}
\overset{6}{\cancel{7}} \overset{1}{2} & 1\ 6 \\
-1\ 6 & +5\ 6 \\
\hline
5\ 6 & 7\ 2
\end{array}
$$

5.
$$
\begin{array}{cc}
\overset{1}{\cancel{2}} \overset{1}{5} & 9 \\
-\ \ 9 & +1\ 6 \\
\hline
1\ 6 & 2\ 5
\end{array}
$$

6.
$$
\begin{array}{cc}
\overset{8}{\cancel{9}} \overset{1}{6} & 4\ 7 \\
-4\ 7 & +4\ 9 \\
\hline
4\ 9 & 9\ 6
\end{array}
$$

7.
$$
\begin{array}{cc}
8\ 9 & 2\ 5 \\
-2\ 5 & +6\ 4 \\
\hline
6\ 4 & 8\ 9
\end{array}
$$

8.
$$
\begin{array}{cc}
9\ 4\ 6 & 6\ 3\ 2 \\
-6\ 3\ 2 & +3\ 1\ 4 \\
\hline
3\ 1\ 4 & 9\ 4\ 6
\end{array}
$$

9.
$$
\begin{array}{cc}
7\ 5\ 1 & 3\ 2\ 0 \\
-3\ 2\ 0 & +4\ 3\ 1 \\
\hline
4\ 3\ 1 & 7\ 5\ 1
\end{array}
$$

10.
$$
\begin{array}{r}
\overset{1\ 1\ 1}{4,8\ 2\ 6} \\
1,4\ 9\ 3 \\
+5,0\ 6\ 6 \\
\hline
1\ 1,3\ 8\ 5
\end{array}
$$

11.
$$
\begin{array}{r}
\overset{1\ \ \ 1}{3,9\ 2\ 6} \\
3,2\ 3\ 8 \\
+1,2\ 2\ 1 \\
\hline
8,3\ 8\ 5
\end{array}
$$

12.
$$
\begin{array}{r}
\overset{2}{9\ 5\ 2} \\
3\ 8\ 1 \\
3\ 8\ 1 \\
+\ \ \ 6\ 3 \\
\hline
1,7\ 7\ 7
\end{array}
$$

13. 56,142; "fifty-six thousand, one hundred forty-two"

14. 224,651

15. 3,000

16. 5,000

17. :21

18. :50

19. 2,452 + 1,079 + 958 = 4,489 people

20. 255 + 125 = 380 dollars
380 − 140 = 240 dollars

Lesson Test 23

1. 4:55

2. 7:15

3. 3:03

4. 6:00

5.
$$
\begin{array}{cc}
\overset{8}{\cancel{9}} \overset{1}{0} & 1\ 3 \\
-1\ 3 & +7\ 7 \\
\hline
7\ 7 & 9\ 0
\end{array}
$$

6.
```
  3 1   1
  44    35
 -35   + 9
   9    44
```

7.
```
  4 1    1
  52     29
 -29    +23
  23     52
```

8.
```
 638    117
-117   +521
 521    638
```

9. $3.78 + $0.55 = $4.33

10. 3 dimes = 30¢
5 nickels = 25¢
30¢ + 25¢ + 16¢ = 71¢ or $0.71

Lesson Test 24

1.
```
  1 9      1 1
  200      38
 - 38    +162
  162      200
```

2.
```
    5       1
  367      149
 -149     +218
  218      367
```

3.
```
  6 4      1 1
  755      286
 -286     +469
  469      755
```

4.
```
  7 7      1 1
  880      94
 - 94    +786
  786      880
```

5.
```
  48      14
 -14     +34
  34      48
```

6.
```
  53      42
 -42     +11
  11      53
```

7. 7:09

8. 3:47

9.
```
   1
  650
 +163
  813
```

10.
```
    1
  274
 +895
 1,169
```

11.
```
    1
 3,039
+5,211
 8,250
```

12. 313 − 194 = 119 pages

13. 10 + 12 + 15 = 37'
37 − 19 = 18'

14. 3,100 + 4,916 = 8,016 miles

Lesson Test 25

1. February
2. fourth
3. twelfth
4. first — Friday
5. second — Tuesday
6. third — Sunday
7. fourth — Wednesday
8. fifth — Monday
9. sixth — Saturday
10. seventh — Thursday
11. 30
12. 31
13. 15
14. 12
15. ||||| ||||| ||||| |||
16. ||||| ||||| ||||| ||||| |||||

17.
```
   2 3
   341
 -  69
   272
```

18.
```
     3
   743
 - 205
   538
```

19.
```
   5 9
   600
 -  35
   565
```

20.
$$
\begin{array}{r}
{\scriptstyle 1}\\
5,158\\
+1,222\\
\hline
6,380
\end{array}
$$

Lesson Test 26

1.
$$
\begin{array}{r}
{\scriptstyle 7\ ^{1}0}\\
7,8\,\cancel{1}\,0\\
-\ \ 681\\
\hline
7,129
\end{array}
\qquad
\begin{array}{r}
{\scriptstyle 1\ 1}\\
681\\
+7,129\\
\hline
7,810
\end{array}
$$

2.
$$
\begin{array}{r}
{\scriptstyle 4\ ^{1}0}\\
5,\cancel{1}\,23\\
-4,882\\
\hline
241
\end{array}
\qquad
\begin{array}{r}
{\scriptstyle 1\ 1}\\
4,882\\
+\ \ 241\\
\hline
5,123
\end{array}
$$

3.
$$
\begin{array}{r}
{\scriptstyle 1}\\
2,\cancel{1}75\\
-1,504\\
\hline
671
\end{array}
\qquad
\begin{array}{r}
{\scriptstyle 1}\\
1,504\\
+\ \ 671\\
\hline
2,175
\end{array}
$$

4.
$$
\begin{array}{r}
{\scriptstyle 8}\\
4\cancel{9}\,3\\
-\ 58\\
\hline
435
\end{array}
\qquad
\begin{array}{r}
{\scriptstyle 1}\\
58\\
+435\\
\hline
493
\end{array}
$$

5.
$$
\begin{array}{r}
{\scriptstyle 7\ ^{1}1}\\
8\cancel{2}\,4\\
-367\\
\hline
457
\end{array}
\qquad
\begin{array}{r}
{\scriptstyle 1\ 1}\\
367\\
+457\\
\hline
824
\end{array}
$$

6.
$$
\begin{array}{r}
{\scriptstyle 3}\\
\cancel{4}\,\cancel{1}\\
-32\\
\hline
9
\end{array}
\qquad
\begin{array}{r}
{\scriptstyle 1}\\
32\\
+\ 9\\
\hline
41
\end{array}
$$

7. 𝍸 𝍸
8. 𝍸 𝍸 𝍸 𝍸 𝍸 𝍸 𝍸 ||
9. 𝍸 𝍸 𝍸 𝍸 𝍸 |||
10. 8 : 40
11. 3 : 20
12. Friday
13. second
14. $43 + 43 + 43 + 43 = 172"$
15. $5 + 8 + 13 = 26$ picked
     $26 - 18 = 8$ left

Lesson Test 27

1.
$$
\begin{array}{r}
{\scriptstyle 5}\\
\$5.\cancel{6}\,7\\
-0.2\,8\\
\hline
\$5.3\,9
\end{array}
$$

2.
$$
\begin{array}{r}
{\scriptstyle 2}\\
\$\cancel{3}.\,19\\
-1.42\\
\hline
\$1.77
\end{array}
$$

3.
$$
\begin{array}{r}
{\scriptstyle 5\ \ 3}\\
6,\cancel{3}\cancel{4}\,0\\
-3,506\\
\hline
2,834
\end{array}
$$

4.
$$
\begin{array}{r}
{\scriptstyle 1\ 1}\\
175\\
+\ 48\\
\hline
223
\end{array}
$$

5.
$$
\begin{array}{r}
{\scriptstyle 1\ 1}\\
\$\ 4.29\\
+\ 6.88\\
\hline
\$11.17
\end{array}
$$

6.
$$
\begin{array}{r}
{\scriptstyle 1\ 1\ 1}\\
7,381\\
2,479\\
+5,630\\
\hline
15,490
\end{array}
$$

7. first ——————January
8. second ⟋April
9. third ⤬May
10. fourth ⤬February
11. fifth ⟍March
12. sixth ——————June
13. 2, 4, 6, 8, 10, 12, 14, 16, 18, 20
14. 𝍸 ||
15. $\$3.50 - \$1.75 = \$1.75$

Lesson Test 28

1.
$$
\begin{array}{r}
{\scriptstyle 3\ ^{1}4}\\
63,\cancel{4}\cancel{5}\,2\\
-41,369\\
\hline
22,083
\end{array}
\qquad
\begin{array}{r}
{\scriptstyle 1\ 1}\\
41,369\\
+22,083\\
\hline
63,452
\end{array}
$$

2.
$$
\begin{array}{r}
{}^{7}{}^{12}\\
75,8\cancel{3}0 \\
-22,536 \\
\hline
53,294
\end{array}
\qquad
\begin{array}{r}
{}^{1}{}^{1}\\
22,536 \\
+53,294 \\
\hline
75,830
\end{array}
$$

3.
$$
\begin{array}{r}
\$14.98 \\
- 6.12 \\
\hline
\$ 8.86
\end{array}
$$

4.
$$
\begin{array}{r}
{}^{1}\\
\$59.\cancel{2}0 \\
-30.15 \\
\hline
\$29.05
\end{array}
$$

5.
$$
\begin{array}{r}
{}^{8}\ {}^{3}\\
\$\cancel{9}\cancel{1}.\cancel{4}2 \\
- 15.17 \\
\hline
\$76.25
\end{array}
$$

6. 26 < 62

7. 2 < 4

8. 9 = 9

9.
$$
\begin{array}{r}
22\\
124 \\
356 \\
279 \\
+451 \\
\hline
1,210
\end{array}
$$

10.
$$
\begin{array}{r}
11\\
729 \\
512 \\
283 \\
+ 45 \\
\hline
1,569
\end{array}
$$

11.
$$
\begin{array}{r}
11\\
109 \\
460 \\
999 \\
+ 31 \\
\hline
1,599
\end{array}
$$

12. Monday

13. second

14. $54.98 – $21.15 = $33.83

15. 19,345 – 18,400 = 945 frogs

Lesson Test 29

1. 15

2. 100°

3.
35°

4. 80

5. 30

6.
$$
\begin{array}{r}
{}^{1}\\
54,0\cancel{2}\cancel{1} \\
-14,015 \\
\hline
40,006
\end{array}
$$

7.
$$
\begin{array}{r}
{}^{6}\ {}^{4}\ {}^{13}\\
\cancel{7},\cancel{2}\cancel{4}\cancel{3} \\
-3,567 \\
\hline
3,676
\end{array}
$$

8.
$$
\begin{array}{r}
1\ 1\\
9,461 \\
+2,743 \\
\hline
12,204
\end{array}
$$

9. 1 : 25

10. 6 : 03

11. 1,799 – 1,732 = 67 years

12. 卌 卌 卌 卌
卌 卌 卌 卌 卌 II

Lesson Test 30

1. February

2. March

3. 25°

4. 40°

5. 5

6. 300°

7.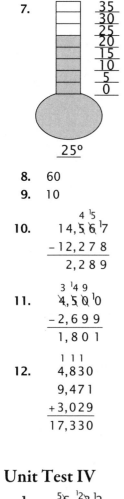

25°

8. 60

9. 10

10.
$$\begin{array}{r} 4\ ^{1}5 \\ 14,5\,\cancel{6}\,^{1}7 \\ -12{,}278 \\ \hline 2{,}289 \end{array}$$

11.
$$\begin{array}{r} 3\ ^{1}4\ 9 \\ 4{,}5\,\cancel{0}\,^{1}0 \\ -2{,}699 \\ \hline 1{,}801 \end{array}$$

12.
$$\begin{array}{r} 1\ 1\ 1 \\ 4{,}830 \\ 9{,}471 \\ +3{,}029 \\ \hline 17{,}330 \end{array}$$

Unit Test IV

1.
$$\begin{array}{r} ^{5}\cancel{6}\ ^{12}\cancel{3}\ ^{1}2 \\ -\ \ 4\ \ 49 \\ \hline 1\ \ 8\ 3 \end{array}$$

2.
$$\begin{array}{r} 7\ ^{4}\cancel{5}\ ^{1}0 \\ -5\ \ 36 \\ \hline 2\ \ 14 \end{array}$$

3.
$$\begin{array}{r} ^{7}\cancel{8}\ ^{1}9 \\ -\ 2\ 63 \\ \hline 5\ 56 \end{array}$$

4.
$$\begin{array}{r} ^{4}\cancel{5},\ ^{1}\cancel{2}\ ^{9}\cancel{0}\ ^{1}8 \\ -\ 1,\ \ 6\ \ 19 \\ \hline 3,\ \ 5\ \ 89 \end{array}$$

5.
$$\begin{array}{r} ^{8}\cancel{9},\ ^{17}\cancel{8},\ ^{12}\cancel{3}\ ^{13}\cancel{4}\ ^{1}2 \\ -\ 6\ \ 8,\ \ 4\ \ 53 \\ \hline 2\ \ 9,\ \ 8\ \ 89 \end{array}$$

6.
$$\begin{array}{r} \$\ ^{6}\cancel{7}\ ^{1}\cancel{2}.\ ^{1}4 \\ -\ \ \ 3\ \ 4.21 \\ \hline 3\ \ 7.93 \end{array}$$

7. 12 : 30

8. 2 : 23

9. 25

10. 9

11. |||| |||| |||| ||||

12. |||| |||| |||| ||||
|||| |||| |

13. Thursday

14. Tuesday

15. fourth

16. twelfth

17. Wednesday

18. 15

19. Monday and Tuesday

20. 20

21. 400°

22.

70°

23. 100

24. 70

Final Test

1. 14 > 7

2. 105 < 125

3. 40

4. 70

5. 200

6. 600

7. 2,000

8. 4,000

9. 2, 4, 6, 8, 10, 12, 14, 16, 18, 20

10. 5, 10, 15, 20, 25, 30, 35, 40, 45, 50

11. 10, 20, 30, 40, 50, 60, 70, 80, 90, 100

12.
$$\begin{array}{r} {}^1 2\,4 \\ +\;4\,6 \\ \hline 7\,0 \end{array}$$

13.
$$\begin{array}{r} {}^1 1\,9\,2 \\ +\;3\,5\,9 \\ \hline 5\,5\,1 \end{array}$$

14.
$$\begin{array}{r} 9\,{}^1 0\,7 \\ +\;1\,6\,8 \\ \hline 1,0\,7\,5 \end{array}$$

15.
$$\begin{array}{r} \$\,{}^1 8.{}^1 9\,2 \\ +\;\;2.4\,9 \\ \hline \$11.4\,1 \end{array}$$

16.
$$\begin{array}{r} {}^1 6,{}^1 4\,7\,4 \\ 7,6\,1\,0 \\ +\;3,6\,8\,5 \\ \hline 1\,7,7\,6\,9 \end{array}$$

17.
$$\begin{array}{r} {}^1 9\,{}^2 6\,8 \\ 1\,4\,5 \\ 2\,0\,3 \\ +\;\;\;7\,5 \\ \hline 1,3\,9\,1 \end{array}$$

18.
$$\begin{array}{r} {}^1 \cancel{2}\,{}^1 3 \\ -\;1\,7 \\ \hline 6 \end{array}$$

19.
$$\begin{array}{r} 1\,{}^0 \cancel{1}\,{}^1 5 \\ -\;\;9\,8 \\ \hline 1\,7 \end{array}$$

20.
$$\begin{array}{r} {}^3 \cancel{4}\,{}^9 \cancel{0}\,{}^1 3 \\ -\;\;2\,1\,5 \\ \hline 1\,8\,8 \end{array}$$

21.
$$\begin{array}{r} {}^6 \cancel{7}\,{}^{10} \cancel{1}\,{}^1 0 \\ -\;\;3\,4\,6 \\ \hline 3\,6\,4 \end{array}$$

22.
$$\begin{array}{r} 5,{}^7 \cancel{8}\,{}^{12} \cancel{3}\,{}^1 4 \\ -\;1,0\;\;5\,7 \\ \hline 4,\;7\;\;7\,7 \end{array}$$

23.
$$\begin{array}{r} {}^7 \cancel{8}\,1,{}^2 \cancel{3}\,{}^{11} \cancel{2}\,{}^1 7 \\ -\;\;4\,5,1\;\;8\,9 \\ \hline 3\,6,1\;\;3\,8 \end{array}$$

24. 276,591

25. 7:23

26. 100°

27. 12+12+12+12 = 48"

28. 14+14+14+14 = 56"

29. 6+11+6+11 = 34'

30. 9+8+6 = 23"

BETA

Symbols and Tables

SYMBOLS

<	less than
>	greater than
=	equals
+	plus
−	minus
¢	cents
$	dollars
°	degrees
'	foot or feet
"	inch or inches

PLACE-VALUE NOTATION

31,452 = 30,000 + 1,000 + 400 + 50 + 2

MONTHS OF THE YEAR

January
February
March
April
May
June
July
August
September
October
November
December

MONEY AND MEASURE

1 penny = 1 cent (1¢ or $0.01)
1 nickel = 5 cents (5¢ or $0.05)
1 dime = 10 cents (10¢ or $0.10)
1 dollar = 100 cents (100¢ or $1.00)
100 centimeters (cm) = 1 meter (m)
12 inches (12" or 12 in) = 1 foot (1' or 1 ft)
5,280 feet (ft) = 1 mile (mi)
1 week = 7 days

DAYS OF THE WEEK

Sunday
Monday
Tuesday
Wednesday
Thursday
Friday
Saturday

LABELS FOR PARTS OF PROBLEMS

Addition

25	addend
+16	addend
41	sum

Subtraction

4 5	minuend
− 2 2	subtrahend
2 3	difference

Glossary

A–D

addend - a number that is added to another

Associative Property - a property that states that the way terms are grouped in an addition expression does not affect the result

base ten - a number system based on ten, also called *decimal system*

cardinal numbers - numbers that indicate quantity (one, two, three, etc.)

circle - a simple closed curve with points that are all the same distance from the center

Commutative Property - a property that states that the order in which numbers are added does not affect the result

decimal point - a dot used to separate whole numbers and fractions; also used to separate dollars and cents

decimal system - a number system based on ten, also called base ten

decompose - to separate a number into parts

difference - the result of subtracting one number from another

E–P

equation - a mathematical statement that uses an equal sign to show that two expressions have the same value

estimate - a close approximation of an actual value

factor - a whole number that multiplies with another to form a product

inequality - a mathematical statement showing that two expressions have different values

minuend - a number from which another is to be subtracted

minus - decrease by subtraction

ordinal numbers - numbers that indicate position (first, second, third, etc.)

perimeter - the distance around a polygon

place value - the position of a digit which indicates its assigned value

plus - increase by addition

R–S

rectangle - a quadrilateral with two pairs of opposite parallel sides and four right angles

regrouping - composing or decomposing groups of ten when adding or subtracting

rounding - replacing a number with another that has approximately the same value but is easier to use

skip counting - counting forward or backward by multiples of a number other than one

square - a quadrilateral in which the four sides are perpendicular and congruent

subtrahend - a number to be subtracted from another

sum - the result of adding numbers

T–Z

triangle - a polygon with three straight sides

unit - the place in a place-value system representing numbers less than the base

unknown - a specific quantity that has not yet been determined, usually represented by a letter

Master Index for General Math

This index lists the levels at which main topics are presented in the instruction manuals for *Primer* through *Zeta*. For more detail, see the description of each level at mathusee.com. (Many of these topics are also reviewed in subsequent student books.)

Beta Index